ROAD & TRACK
ON
AUSTIN HEALEY
1953-1970

Reprinted From
Road & Track Magazine

ISBN 0 946 489 106

Published By
Brooklands Books with permission of Road & Track
Printed in Hong Kong

Titles in this series

Road & Track on Austin Healey 1953-1970
Road & Track on Corvette 1953-1967
Road & Track on Corvette 1968-1982
Road & Track on Ferrari 1968-1974
Road & Track on Ferrari 1975-1981
Road & Track on Fiat Sports Cars 1968-1981
Road & Track on Jaguar 1974-1982
Road & Track on Lamborghini 1964-1982
Road & Track on Lotus 1972-1983
Road & Track on Maserati 1952-1974
Road & Track on Maserati 1975-1983
Road & Track on Mercedes Sports & GT Cars 1970-1980
Road & Track on Porsche 1972-1975
Road & Track on Porsche 1975-1978
Road & Track on Porsche 1979-1982

Titles in preparation will cover:–

Alfa Romeo, BMW, Triumph, MG, Rolls Royce & Bentley, plus further titles on Ferrari, Porsche, Jaguar and Lotus

Distributed By

Road & Track
1499 Monrovia,
Newport Beach,
California 92651, U.S.A.

Brooklands Books Distribution Ltd.,
Holmerise, Seven Hills Road,
Cobham, Surrey KT11 1ES,
England

CONTENTS

We are frequently asked for copies of out of print Road Tests and other articles that have appeared in Road & Track. To satisfy this need we are producing a series of books that will include, as nearly as possible, all the important information on one make or subject for a given period.

It is our hope that these collections of articles will give an overview that will be of value to historians, restorers and potential buyers, as well as to present owners of these automobiles.

THE HEALEY "HUNDRED"

by John Bond

AUSTIN-HEALEY "100" SPECIFICATIONS

General: Wheelbase 90 in.; Tread, front 48.75 in., rear, 49.5 in.; tire size 5.90 x 15; curb weight, 1952 lbs.

Engine: Austin A-90, 4 cyl.; 3 7/16 x 4⅜; 162.2 cu in. (2660 cc); 90 bhp at 4000 rpm. Torque 144 ft/lbs. at 2000 rpm. Twin SU carburetors.

Clutch: Borg and Beck S.D.P., 10 in.

Transmission: Austin 4-speeds forward, synchromesh on 2nd, 3rd, and 4th. Central remote control lever.

Overdrive: Laycock De Normanville, manual control, .756 ratio.

Ratios	Trans.	Direct	Overdrive
4	1.00	4.12	3.12
3	1.42	5.83	4.31
3	2.25	9.28	7.01
1	3.60	14.83	11.21

Rear axle: Austin ¾ floating, spiral bevel gears, 4.125 to 1 ratio.

Suspension: I.F.S. by wishbones and coil springs, rate 79 lbs/in. Rear, semi-elliptic leaf springs, rate 92 lbs/in.

Brakes: Girling two leading shoe hydraulic with 10 x 1.75 in. drums. Area 135 sq. in.

Wheels: Rudge Whitworth type, knock off wire. 15 in. rims.

Equipment: 12 volt electrical system with two 6 volt batteries. Speedometer, tachometer, ammeter, oil pressure, and water temperature gages.

Performance: top speed 110 mph. Acceleration from zero to 60 mph, 10.5 secs.; to 80 mph, 20 secs.; to 100 mph, 37.2 secs. Standing ¼ mile, 18 secs.; standing kilometer, 32.6 secs.

Even with top erected the Healey 100 presents a neat appearance. Overall height is 49 in.

Any new sports car with the clean modern lines of the new Austin-Healey Hundred would get attention. A new sports car with a top speed of 110 mph and acceleration to match would deserve attention. But when you combine these qualities of appearance and real performance with a price of just over $3000, the tremendous interest aroused by this new Healey is understandable.

While Mr. Donald Healey's firm is relatively new in the manufacturing field (since 1946), Mr. Healey has had an active and successful background in the sports car field —both as a competitor and an engineer. If one were to list outstanding characteristics of the various postwar models of Healey cars, it would certainly include high performance and that most difficult combination —excellent handling with good riding qualities. The new "100" uses a normal Austin A-90 engine (90 bhp) in a very compact chassis, resulting in only 22 lbs of weight for each horsepower. The *Road and Track* high gear performance index is 69.7, or better than any stock American sedan. The top speed in 4th gear is given as 92 mph, and using the overdrive (standard equipment), the car was clocked by the Belgian Automobile Club at 111.7 mph.

The Austin-Healey is a joint venture, having the official backing of the British Austin Company, and it will be sold and serviced by Austin dealers throughout the world. The Austin Company (or its suppliers) are furnishing the engine, clutch, transmission, rear axle, front suspension, brakes, and steering. Healey supplies the frame, body, and the assembly facilities.

Frame . . .

The frame is an interesting structure made up of a pair of 3 inch square box section rails laid parallel to each other and to the ground—and only 17 inches apart. These rails are so low that the lower wishbone is above the rail, and they run straight back, dipping slightly to pass under the rear axle. An X-member is used, completing what is

Chrome plated wire wheels are genuine knock off type. Functional streamlining gives an honest top speed of over 110 mph.

virtually a backbone type frame. The rear cross member carries the spring shackles while an outrigger on each side carries the spring eyes and supports the body and floor structure just ahead of the rear wheels.

The cowl is an aircraft style structure tied to the frame by outriggers at the floor and by box section members diagonally from near the top of the cowl down to the front suspension brackets. These diagonals, incidentally, preclude any possibility of using a V-type engine in this chassis.

Suspension . . .

The suspension system is by the modern convention of coil springs in front, leaf springs in rear, and the ride rates are in accordance with American practice, with the front end "softer" than the rear. Since no cars are yet available in this country, we have to accept the words of a British test report—"The suspension is first class, and one is at no time conscious of the short wheelbase. I have not watched this car being cornered fast by another driver, but from behind the wheel there is no apparent roll under even the most extreme conditions." (John Bolster, for *Autosport*, Oct. 24, 1952)

Engine . . .

As shown in the data section, the engine is roughly equivalent to one-half of a big American ohv V-8. This big four proved its stamina in 1949 when it ran 10,000 miles at an average speed of 70 mph on the Indianapolis track. Our engine section shows the Austin A-70 engine which is identical to the A-90 powerplant except for the cylinder bore, pistons and carburetion. The A-90 engine also enjoys parts interchangeability with the Austin A-135 six cylinder engine.

The A-70/90 engine was designed and developed by the Austin Company during the war for military purposes, and first appeared as the Austin "16". The A-90 version, with its 5/16 inch larger bore, was achieved by modification of water jacket coring in the block, but uses the same crank, rods, valves, etc. as the A-70.

There are 3 main bearings, 2.48 inches in diameter, with 2.00 inch diameter crankpins. Connecting rods are 8.19 inches long and are slightly offset. Perhaps the most interesting engine feature is the careful design of ports and combustion chambers—the work of the consulting firm of Westlake and Company, Ltd. The result of apparently minor changes is very good engine performance with moderate valve lift and conservative valve timing (5°-45°-40°-10°). Note how this engine develops the same bhp as the 161 cu in. Willys 6 at 400 less rpm, has 10 more ft/lbs of torque at 2000 rpm, with a lower compression ratio (7.5 to 1).

Despite the engines' 4.38 inch stroke, the cruising speed in overdrive is 83.5 mph at a piston speed of 2500 fpm.

Rear Axle . . .

The Austin-Healey rear axle is one of the few still in production built on the 3/4 floating principle. This gives an additional safety factor and also allows very large rear wheel bearings of the double row ball type. Axle shafts have upset flanges to eliminate the inherent weakness of a taper and key design. The differential proper is unconventional in that the gears are not hypoid, but instead are spiral bevels. The pinion gear is supported by a pair of pre-loaded Timken bearings, while the carrier bearings are very large angular contact ball bearings.

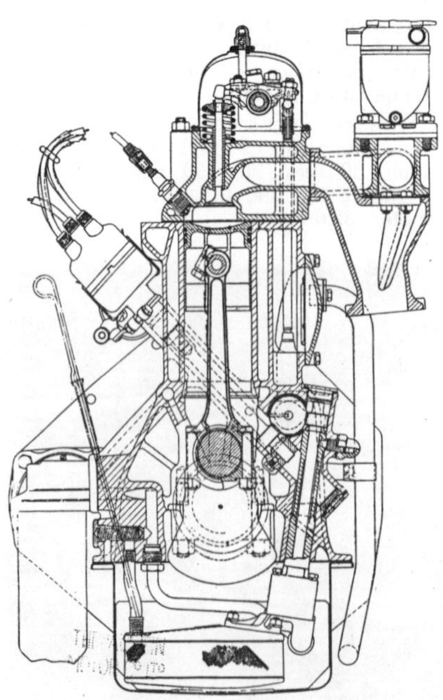

Section view of sturdy Austin engine. (*Drawing courtesy of Automobile Engineer*)

Transmission . . .

The clutch is larger than the A-70/90, but the transmission is identical except for a central control lever. This transmission appears well designed, but the typical British creeper low is superfluous. Second gear is adequate for all starts in the 1952 lb. Healey (the A-90 curb weight is 3000 lbs.). Second, third, and fourth gears are synchronized by spring loaded cones which have many advantages over the usual "balking" rings—more durable, faster shifting.

The Laycock De Normanville overdrive (standard equipment) should not be confused with the American Borg-Warner unit. The L. De N. overdrive is controlled manually through friction clutch surfaces. It can be shifted in or out at any speed, and at any time. Smooth shifts are very easy to make; a slight release of the accelerator on up shifts and full throttle for down shifts is the approved technique—but not essential. The actual overdrive ratio is much more sensible than the BW unit. Using the 4.125 axle ratio of the "100", the L. De N. overdrive changes the overall ratio to 3.12 whereas a BW unit would drop the ratio to 2.89. The L. De N. overdrive is a "closer" ratio, which is desirable in a high performance sports car.

Body . . .

All body structure is welded to the chassis frame for added stiffening. Fenders and doors are made of sheet steel, but all other body panels are 16 gage aluminum. The curved glass windshield is positioned by links which are normally concealed but allow the windshield to move forward on the cowl and drop down at the top, giving a rakish angle.

The top stows completely out of sight when not in use, can be easily handled by one person, and is made of an entirely new plastic material. Side curtains are made of formed clear plastic material (to follow the windshield frame shape) with metal frames. Unusual, too, is the provision for 10 cubic feet of storage space in the rear deck.

Rudge-Whitworth type wire wheels with knock off hubs are standard equipment and carry 5.90 x 15 tires.

All things considered, this new car will, when it makes its appearance, make everyone take notice. The line forms behind me•

THE AUSTIN-HEALEY AT BONNEVILLE

Shade is a priceless luxury on the Utah salt flats where temperatures often exceed 100° F.

Geoffrey Healey consults with an AAA official. This is the stock car used at Bonneville.

Photographs by Borgeson

During the International Record Trials held at Bonneville in September, a pair of Austin-Healeys set-up over 170 new records for speed and endurance.

The most spectacular new record established is for modified production sports cars—142.636 mph over the measured mile. Thus a class D car (2 to 3 litres) has exceeded the previous class C (3 to 5 litres) mark of 141.846 mph. Driven by Donald Healey, George Eston, Bill Spear and John Benett, the same car later set up new records up to 12 hours. The car itself carried a special underpan, a cockpit cover and a streamlined headrest. It ran on gasoline and obviously puts out more than the stock 90 bhp.

Less spectacular perhaps, but even more significant, are the new production stock car records. This car (illustrated) finally cracked the 16 year old American unlimited record of 107.66 mph held by a supercharged Cord, by clocking 109.243 mph for 5 miles. Driven by Benett, Cooper and Jackson-Moore, the stock car continued on for 30 hrs., setting up new records all the way. The 24 hour mark for International Class D racing cars was 98.16 mph, held by a Citroen Special since 1934. The Austin-Healey averaged 104.32 mph for the 24 hour distance. ●

Austin-Healey Records

Modified Sports Car	
1 mile (2-way avg.)	142.636 mph
6 hours	123.750 mph
12 hours	122.913 mph
Production Sports Cars	
5 miles	109.243 mph
10 miles	108.788 mph
1 hour	105.180 mph
12 hours	104.600 mph
24 hours	104.320 mph
30 hours	104.000 mph

Greater performance than ever!

presents the new '56

Austin-Healey 100

Check these improved features . . .

New, close ratio, four speed transmission with overdrive . . . New front suspension incorporating taper roller bearings for longer life at high speeds . . . Larger, more powerful brakes with anti-fade linings.

Arrange to test-drive the new Austin-Healey 100 today.

Represented in the United States by **hambro** AUTOMOTIVE CORPORATION
27 W. 57th ST., NEW YORK 19, N. Y.
Sold and serviced through a nationwide network of distributors and dealers.

The Austin-Healey 100 is a product of THE BRITISH MOTOR CORPORATION, LTD. Makers of MG, AUSTIN-HEALEY, MAGNETTE, MORRIS, AUSTIN AND RILEY CARS.

Road Testing the Austin-Healey

photographs by
Rolofson

The Austin-Healey 100 is a perfect example of fairly "ordinary" components assembled together in such an ingenious manner that the results are almost unbelievable. Even when the "100" was first announced, late in 1952, it was bound to be well received on the score of appearance and price alone. Performance data on an early prototype model were quickly published and again it was obvious that the new car would "go" in keeping with its rakish appearance, even allowing for the usual performance depreciation of production models which inevitably become heavier and slower.

But the Austin-Healey was slow to get into quantity production and some of the early enthusiasm seemed to wane. When the first few cars were delivered, late in 1953, rumors began to circulate. We drove three different early models, talked to dozens of owners (later) and have yet to find any verification of serious "bugs."

The biggest mis-conception concerning the Austin-Healey is its true top speed, *in full touring trim!* In much (very much) modified form it has achieved 142.63 mph on the Bonneville Salt Flats. That was with 135 bhp and a certain amount of body fairing. In slightly modified form, developing just over 100 bhp, the Le Mans model attained 119.12 mph on the famous circuit, in competition. The following table shows a comparison of our timed top speed runs with two British magazine road tests, and speaks for itself.

Magazine	R & T	"A"	"B"
Average	102.3	103	101.5
Best run	106.0	108	107.2*
Gear ratio	3.12	3.12	2.78
Test weight	2470	2500	2464

* *with windshield and top folded down*

From this table, it can be reliably stated that the true maximum speed of a production A-H must lie between 101.5 and 103 mph with windshield erect, top and side curtains in position. However, we did not make any runs with the unique windshield in its retracted position, or with the top and curtains removed. British road tests indicate that in this modified trim the top speed is increased by 4.5 to 8 mph.

No further remarks on the top speed are really necessary, but a genuine 100 plus with only 90 bhp (and in a roadster at that) is a truly remarkable figure. Not only that, but to put it quite frankly, the A-H is one of the very few cars which we have driven at this speed without adding a few "gray hairs." Despite a very nasty cross-wind the four timed runs showed-up the faultless directional stability of the car—it holds a straight line without the slightest difficulty and always felt perfectly safe.

The excellent handling characteristics of the A-H are not confined to stability at high speed. In fact, our jury of testers were unanimous in rating the "100" as the best all-around handling car encountered this year. The ride is at least as good as any sports car we've tested in the past year, cornering is very flat and controllable, there is a small amount of understeer, a good feel of the road with no kick-back,—all to be added to the previously mentioned high speed stability. The steering itself is exceptionally light, yet requires only 2.5 turns, lock to lock. The only fault we could find in this department was a slight backlash in the steering, which certainly did not seem to hurt the controllability. This particular test car had 10,000 miles on the odometer and the steering gear had never been adjusted in that time.

The acceleration performance data and graph require some explanation. In the first place the clutch gave us a lot of trouble, during the acceleration testing. For ordinary traffic driving the clutch action is perfectly normal—smooth in engagement and requiring no finesse whatsoever. But as soon as the best possible "take-off" was attempted, the action became extremely fierce. The clutch comes in with a bang, there is a lunge forward accompanied by a chirp from the tires—but then the car stumbles momentarily. The best technique seems to require wide open throttle the instant the clutch grabs hold. Very rapid starts can be made by using only 1000 rpm at the "line" and giving full throttle as soon as the car has moved two or three feet. We suspect that this clutch idiosyncracy was actually a manifestation of rear spring wind-up.

The 3-speeds with overdrive presented another problem. The shift pattern is non-standard, which seems stupid, but it was quickly learned and actually proved no great inconvenience except that occasionally the shift to second is missed by going up a blind alley, in the "gate." But with five speeds available, it was necessary to experiment with various combinations to achieve the best possible times to 60, 70, 80 mph, etc. Our acceleration graph shows that we found it best to actually omit using high gear for these tests. We also discovered that the hydraulically actuated clutch of L.deN. overdrive will begin to slip after three or four successive snap shifts from 2nd to 2nd od under full throttle. However, it only happened once. Syncromesh on first gear was a great boon, but this particular car did not allow really fast shifts to be made with out a slight clicking as the dogs engaged. The tachometer, incidently is "red-lined" at 4800 rpm, a piston speed of 3500 ft/min. We used 4500 rpm as a limit, but once reached 5000 rpm (about 90 mph) so quickly in high gear that we immediately cancelled all thought of a timed run in that ratio. The speedometer wabbles a bit and indicated 115/118 during the best timed run in overdrive.

A big four-cylinder engine is never as quiet and smooth as a six or an eight, but the Austin A-90 is surprisingly good in this department. It would be extremely difficult to say whether the A-H has 4, or 6, or 8 cylinders, if you didn't know. There is an almost imperceptible vibration period at around a steady 65 mph, indicated, in overdrive. Like all in-line fours there is some vibration when decelerating with the foot off the accelerator pedal. Considering that there is very little sound-deadening or insulation in the A-H, the engine is quiet, the overdrive admittedly contributing its share to this effect. The only serious objections to this powerplant was that it smoked noticeably and showed, at times, an extreme reluctance to stop — even with the ignition turned off and the engine temperature gage reading 70° C. Rumors to the contrary, the engine temperature never went over 80° F at any time and although the transmission tunnel does get warm, the ash-tray (in the tunnel) does not get uncomfortably hot.

As is our usual practice, we made no brake test, preferring not to return cars loaned to us with scorched linings and cracked drums. But the well ventilated, 11 inch drums on the A-H give every indication of being quite adequate for a car of such moderate weight. A ratio of only 14.8 lbs (curb) for each sq in. of lining area assures adequate lining life and we found the brakes to be eminently satisfactory in every way. The only braking fault was due to the tires—they seemed to squeal at every stop when inflated at 24 psi, an annoyance only partially alleviated by using higher pressures. On this, a reasonable excuse might be offered on the basis that the deceptive performance and good brakes were contributing factors, or in other words, we were over-driving without realizing the fact.

The astounding efficiency of the four cylinder engine coupled to an overdrive is amply proven by the fuel consumption data. Highway driving at 55/60 mph gave 27 miles per gallon. Breezing along an indicated 80 mph most of the time, nearly 24 mpg. The performance testing required 63 miles and even then the mileage dropped only to 21 mpg. The overall figure of 319 miles of varied driving came out at exactly 24.35 mpg, a very creditable figure.

How much better the Austin-Healey will perform with the optional "Le Mans" engine modifications is pure guess work. The factory says that the bhp becomes 100 at 4500 rpm. This appears to a very conservative figure for a camshaft with higher lift and more duration will certainly allow more than the stock 7.5 to 1 compression ratio. Add to that a pair of larger carburetors and the true bhp should be well in excess of 100. A fair estimate would place the zero to 60 mph time close to 10.0 seconds, the standing quarter mile at just over 17.0 seconds. What the top speed would be depends of course on whether the windshield is erect, folded or removed. As stated earlier, the 1953 Le Mans cars recorded 119 mph but this was with windshield removed and running special gear ratios. In any event, the stock Austin-Healey gives a very remarkable performance which requires no apology.

Considering body and comfort features the suspension design gives a front end mo-

Side trip up narrow canyon provided opportunity for testing adhesion on gravel road.

Curved windshield posts add interest to the frontal aspect of the Austin-Healey 100.

Side view shows why the car is so popular.

The "100" is graceful from any angle, has good luggage capacity and adequate bumpers.

tion of about 76 oscillations per minute, with the rear frequency slightly higher at 86 opm. Noting the fact that a lower opm figure means a "softer" ride, compare this table of opm.

Make	A-H	Corvette	Allard PB
	Oscillations / Minute		
Front	76	71	65
Rear	86	84	92
Average	81	78	79

Sitting in the car for the first time there is an impression of being swallowed, since the seats are extremely low. The two individual seats are very comfortable and being separated by a wide tunnel there is ample elbow room, further assisted by large holes in the inner door panels. The test car had been fitted with a two-inch extension on the clutch pedal pad and this seemed to be a worthwhile improvement. The steering wheel is adjustable via a sliding spline but the lock-nut is recessed into the instrument panel so that it is very inconvenient. This car also had the early signal-proof one-piece plexiglass side curtains and some form of ventilation would be appreciated. Later models have a more conventional type of side curtain with signalling flaps. The top proved very simple and quick to operate when a two-man team learned the proper sequence.

Summed up, the Austin-Healey 100 is an exceptionally good all-around sports car—one which will satisfy the desires of a large number of enthusiasts who want moderate price, good looks, high performance and reliability. ●

ROAD AND TRACK ROAD TEST NO. F-7-54

AUSTIN-HEALEY 100

SPECIFICATIONS

List price	$2985
Wheelbase	90 in.
Tread, front	48.8 in.
rear	49.5 in.
Tire size	5.90-15
Curb weight	2150 lbs
distribution	49/51
Test weight	2470 lbs
Engine	4-cyl.
Valves	ohv
Bore & stroke	3.44x4.38
Displacement	162.2 cu in. (2660 cc)
Compression ratio	7.50
Horsepower	90
peaking speed	4000
equivalent mph	95.6
Torque, ft/lbs	144
peaking speed	2500
equivalent mph	59.7
Mph per 1000 rpm	23.9
Mph at 2500 fpm piston speed	82.0
Gear ratios (overall)	
overdrive	3.12
high	4.12
2nd od	4.42
2nd	5.85
1st	9.28
R & T perf. factor (high)	65.1

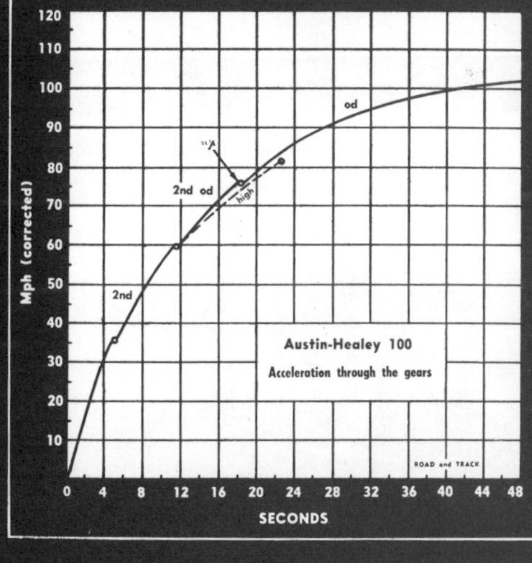

PERFORMANCE

Top speed (avg.)	102.3
fastest one way	106.0
Max speeds in gears—	
high	87
2nd od	81
2nd	61
1st	38
Shift points from—	
2nd od	76 mph
2nd	60 mph
1st	36 mph
Mileage	20/27 mpg

TAPLEY READINGS

Gear	Lbs/ton	at	Mph
1st	590	at	22
2nd	450	at	33
2nd od	340	at	42
high	320	at	45
od	240	at	55

COASTING

(wind and rolling resistance)

92 lbs/ton	at	60 mph
42 lbs/ton	at	30 mph
30 lbs/ton	at	10 mph

ACCELERATION

0-30 mph	4.0
0-40 mph	6.2
0-50 mph	8.3
0-60 mph	11.7
0-70 mph	15.7
0-80 mph	20.8
0-90 mph	27.4
Standing start ¼ mile—	
average	18.1 secs.
best	17.9 secs.

SPEEDO ERROR

Indicated	Actual
10	12.4
20	20.1
30	28.2
40	37.2
50	47.0
60	56.4
70	65.9
80	75.5
90	84.5

Testing the "100S"

a competition car at a reasonable price

photography: **Rolofson**

THERE ARE two "schools of thought" on sports cars—those who want as much comfort as can be had without sacrificing the need for the lowest possible weight, and the extremist group who will literally give up anything and everything for the goal of maximum performance with minimum weight.

The Austin-Healy 100-S competition model falls into the second category, but at the same time there are many surprising creature comforts and it would not be a difficult task to add the stock model's windshield and top, though they are not cataloged options. We mention this because, although the 100-S is designed primarily to suit the needs of the man who wants to race, it is surprisingly easy to drive—with none of the traffic-temperament usually found in "all-out" competition type sports cars. The engine is docile, the clutch operates easily and the brakes are light. Nor is there the usual "tinny" noise and rattle of most pure racing machines.

The tractability of this car becomes all the more remarkable when we explain that Gough Industries (West Coast Distributors for all B.M.C. products) cleverly contrived to change the axle to the standard 2.92-to-one ratio. Our test car was in fact the same machine which ran at Pebble Beach, where a 3.66 axle ratio was used (other options are 2.69 and 4.12). As our data panel proves, the acceleration times are exceptionally good, especially when it is borne in mind that the "high" gears used during our tests were selected to give the best possible top speed—albeit at some sacrifice in acceleration.

While the results of the timed runs for maximum speed were somewhat below our expectations (as well as the local distributor's) we believe they are truly representative of a typical car in good tune. The engine "peaked-out" from a standing start in about 1½ miles and there was no sign of any more "revs" as we approached the timing strip. The tachometer held 4600 rpm steadily on the two best runs in opposite directions. In our opinion the 42 extra horses supplied in the "S" model, as compared to the stock 90 bhp "100" machine, should add 15 mph to the top speed in round numbers —which is just about what happened.

The drag figures at 60 mph are 121 lbs for the 100-S and 114 lbs for our July, 1954 test on the "100", a reduction of 6% in favor of the full-windshield, top-erected, side-curtains-installed condition. How did Donald Healey manage 143.13 mph on the Salt Flats? That car had a cockpit cover, an underpan and no windshield, plus the advantage of factory tuning—which figures!

During the acceleration tests it was extremely easy to get too much wheelspin but we enjoyed the services of Bill Pringle, a competition sports car driver of considerable experience. He had no trouble at all in making good fast starts, and subsequent experience with this machine (and another nearly new one) showed that the one vice of the 90 bhp A-H, rear-end bottoming, is not present in the 100-S during standing starts.

As might be expected the car is a joy to drive. The A-H has always been noteworthy in the good-handling department and the "100-S" is equally good for the simple reason that the "100" couldn't be much improved. The ride is noticeably firmer and cornering roll is either non-existent or so small as to be non-detectable. But it is the beautiful close-ratio 4-speed gearbox that really "makes" this machine, both in town and on the highway. You can exceed the usual city speed limit in first gear, while

Our test car in action at the 1955 Pebble Beach event, Bill Pringle driving in the rain.

Manifolds have been moved to the R.H. side.

R.H.D. has advantages for competition work.

second will show 70 mph before you can catch your breath. Out on an open road, sprinkled with politely-spaced Diesels we frequently used second gear for passing, just to hear the exhaust noise as we swept by. Over 100 mph can be seen in third gear (though the speedometer error was scandalous), and this ratio is normally as high as one goes in town. However the four-cylinder engine is remarkably willing and it will lug at 20 mph in high gear if you insist.

The powerplant also deserves mention for it incorporates a number of important modifications, aside from those designed to increase the bhp by 46.6% and the torque by 16.6%. Outstanding on the list are: a nitride-hardened crankshaft, tri-metal bearings, special con-rods without pinch-bolts, solid-skirt pistons and an oilcooler. Every one of these features is a durability item which should contribute enormously to the stamina of the engine, red-lined incidentally at 4800 rpm equivalent to a piston speed of 3500 fpm.

The Dunlop disc brakes (generally called spot-types in this country) need no special comment for the reason that they work admirably and aside from some tendency to squeak they were unobtrusive in action. Stopping distance being a function of tire adhesion there is no reason to believe that any brake can defy the laws of friction, but certainly these brakes are the best type yet developed in their ability to stand up to road-racing usage without fear of fade.

We can find little or no fault with the Austin-Healey 100-S. It offers a genuine competition machine in the not-too-competitive class D at a very fair price. ●

The huge aluminum gas tank holds 24 gallons.

ROAD AND TRACK ROAD TEST NO. F-11-55

AUSTIN-HEALEY "100S"

SPECIFICATIONS

List price	$4995
Wheelbase	90.0 in.
Tread, front	49.6 in.
rear	50.8 in.
Tire Size	5.90-15
Curb weight	2115 lbs.
distribution	51/49
Test weight	2310 lbs.
Engine	4-cyl.
Valves	pohv
Bore & stroke	3.44 x 4.38 in.
Displacement	162.2 cu. in.
Compression ratio	8.30
Horsepower	132
peaking speed	4700
equivalent mph	125
Torque, ft/lbs	168
peaking speed	2500
equivalent mph	66.5
Mph per 1000 rpm	26.6
Mph at 2500 fpm	82.0
Gear ratios (overall)	
4th	2.92
3rd	3.88
2nd	5.57
1st	8.98
R&T performance factor	52.4

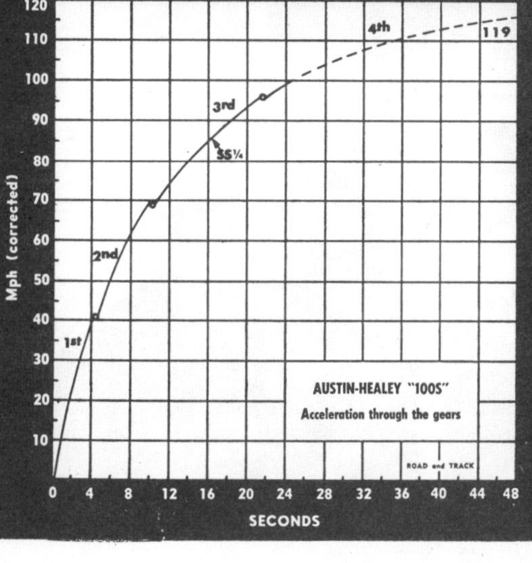

PERFORMANCE

Top speed (avg)	118.6
best run	120.2
Max. speeds in gears—	
3rd (4800)	96
2nd (4800)	67
1st (4800)	42
Shift points from— same as above	
Mileage	not taken

ACCELERATION

0-30 mph	2.8 secs
0-40 mph	4.1 secs
0-50 mph	5.6 secs
0-60 mph	7.8 secs
0-70 mph	10.4 secs
0-80 mph	13.9 secs
0-90 mph	18.2 secs
0-100 mph	24.4 secs
Standing ¼ mile—	
average	16.1 secs
best	15.9 secs

TAPLEY READINGS

Gear	Lbs/ton		Mph	Grade
1st	off scale		—	—
2nd	535	at	43	28 %
3rd	390	at	55	15.5 %
4th	270	at	65)	13.5 %

Total drag at 60 mph, 121 lbs.

SPEEDO ERROR

Indicated	Actual
30	29.9
40	38.2
50	46.1
60	54.5
70	63.9
80	72.0
90	81.3
100	91.0
132	119

Road Test:
THE AUSTIN-HEALEY 100-M
improved performance thanks to the "LeMans kit"

photography
Ralph Poole

Ever since the "LeMans" kit was announced as optional equipment for the Austin-Healey "100," we have been besieged by inquiries—what is included, what is the cost, how much is performance improved, etc.—and some answers are about due.

For 1956, the Austin company have announced a third model for the Healey line-up, the 100-M. For $290 more than the price of a standard 100 model, one gets the following extras, factory installed.

1. An 8.10-to-1 compression ratio obtained by a steel head gasket.
2. A high-lift camshaft.
3. Special valve springs, cups and seats.
4. Two 1¾" S. U. carburetors and necessary attaching parts.
5. A cold-air box for the carburetors.
6. A special advance-curve distributor.
7. A new 4-speed gearbox (overdrive is also standard equipment).

The above items give an engine output of 110 bhp at 4500. Thus the 100M falls between the 90 bhp (at 4000) of the standard 100 model and the 132 bhp at 4700 of the strictily-for-competition 100-S model tested last September. The performance, as might be expected also falls about midway between models, per the following data.

Model	100	100M	100S
top speed	102	109	119
0-60 mph	11.7	9.6	7.8
SS ¼ mile	18.1	17.4	16.1

The 100 model was tested in our July, 1954, issue and all data applies to the cars in "showroom" condition; running with top and side curtains installed except on the 100 S which has neither. In addition, the above performance data can be duplicated by the average owner, driving a car in good tune and making brisk but not brutal gear changes. Furthermore, the performance figures given in the tabulation are all the more remarkable when we take note of the fact that this car weighed 235 lbs more than our earlier 100 model test car.

The 100 M has the stiffer Le Mans type rear springs, front shocks and anti-roll bar. Consequently it rides a little firmer than the normal 100 model and in our opinion suffers somewhat in comparison—because of the really excellent ride in the 100. However, there is no denying that the 100 M suspension is better suited for competition work, and the tendency to bottom (at the rear) on fast starts from a standstill is greatly reduced. The test car was also equipped with the latest type Dunlop Road Speed tires which have sipes (small cuts) in the tread. These were a tremendous improvement over earlier equipment, not only on dry roads, but more especially in the wet. Adhesion on slippery curves is almost uncanny.

The new transmission unit is the same as supplied on the 100 S model, but with overdrive added. This is actually a modified Austin "Princess" unit, designed to withstand over 200 ft-lbs of torque. Durability should be almost unlimited.

Our test car was brought over for the Nassau races by Donald Healey, and it garnered a 9th place in class D, beaten by one other A-H and a flock of Ferraris. The driver was Roy Jackson-Moore, who then drove the car to California and was also at the wheel during our performance tests. Despite nearly 5000 miles on the clock, the car was in good condition and got only minor attention (plugs, points and tappets) before the test. Roy reports that at Nassau, with windshield removed, he indicated 4400 rpm on the long straight, equivalent to about 116 mph, with tire expansion. The best we could get during the timed high speed runs was 4150 rpm, due to the extra drag of top and windshield.

As shown in the data panel, the 3rd-od and 4th-direct ratios are nearly identical, consequently all acceleration tests were made by flipping the overdrive switch from direct to od when 4500 rpm was attained in 3rd gear. This gives a quicker shift and better times than using 4th-direct. The engine is red-lined at 4800 rpm, but there was no improvement in acceleration times by exceeding the 4500 rpm point in each gear. The overdrive unit is available with 3 ratio options: .756, .778, or .820. This car was equipped with the .778 ratio which, when multiplied by the rear axle ratio of 4.10, gives an overall ratio of 3.1898. As now supplied, the overdrive is operable at any speed in any gear—in effect 8 speeds forward, 2 in reverse. However, most owners will prefer to use the overdrive only on the open highway as an extra "5th speed" for cruising. Direct drive (4.10) is a very useful gear for city driving, thanks to the very good lugging power of the large, four-cylinder engine. Top speed in direct drive is limited by the red-line—though the engine will run well beyond 5000 rpm in this ratio if allowed to do so.

The "Le Mans" kit is certainly designed to give a worthwhile performance gain, yet still retain the low speed pulling power (torque) and ruggedness of the large, four-cylinder engine. However, the valve timing of the 100 M is still extremely conservative at only 10° overlap, and for those who are not interested in production car competition a reground camshaft can be used to good advantage. Such a change is described this month in "Tune Up Clinic" on page 42.

Other items which are listed by the Austin Company as options include an 18.7 or a 31 gallon fuel tank, Alfin brake drums and aero screens, all at extra cost. A heater is still supplied and included in the price.

The 100 M can be identified at a glance by the hood louvers and leather hood strap. In all other appearance details the car is identical to previous models, though the overdrive switch and the ignition lock have been interchanged. The seats are comfortable, the driving position is just right and there is ample storage space for odds and ends. In fact, the popularity of the Austin-Healey is due to only two things: (1) it's a genuine sports car and (2) it is reliable and trouble free. The 100 M will enhance that reputation.

ROAD & TRACK ROAD TEST NO. F-3-56

AUSTIN-HEALEY 100M

SPECIFICATIONS

List price	$3275
Wheelbase	90 in.
Tread, front	49.0 in.
rear	50.8 in.
Tire size	5.90-15
Curb weight	2385
distribution	48/52
Test weight	2700
Engine	4 cyl.
Valves	pohv
Bore & stroke	3.44 x 4.38 in.
Displacement	2660cc
Compression ratio	8.10
Horsepower	110
peaking speed	4500
equivalent mph	105
Torque, ft/lbs (est.)	160
peaking speed	2500
equivalent mph	58
Mph per 1000 rpm (od)	23.4
Mph at 2500 fpm (od)	80.1
Gear ratios (overall)	
4th (od)	3.19
4th (high)	4.10
3rd (od)	4.24
3rd	5.45
2nd	7.84
1st	12.6
R & T high gear perf. factor	65.5

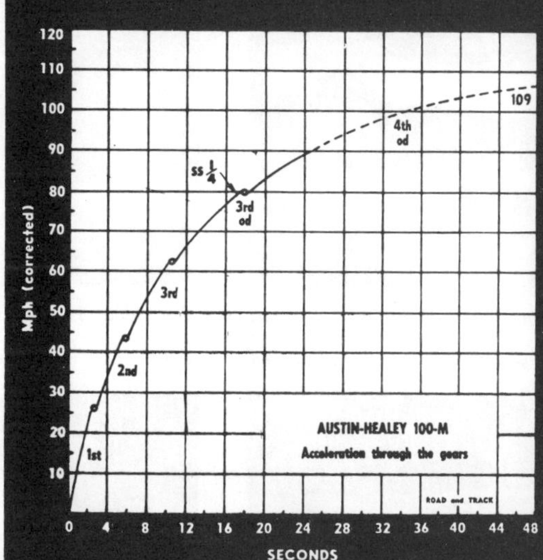

PERFORMANCE

Timed top speed	109
Max. speeds in gears	
4th (4800)	87
3rd (4800)	67
2nd (4800)	46
1st (4800)	28
Shift points from	
3rd od (4500)	80
3rd (4500)	62
2nd (4500)	43
1st (4500)	26
Mileage range	22/28 mpg

ACCELERATION

0-30	3.4 secs.
0-40	4.7 secs.
0-50	7.2 secs.
0-60	9.6 secs.
0-70	13.1 secs.
0-80	17.9 secs.
0-90	24.3 secs.
standing ¼ mile	17.4 secs.

TAPLEY READINGS

Gear	Lbs/ton	Mph	Grade
1st	600	23	32%
2nd	530	35	28%
3rd	430	44	22%
od	330	52	17%
4th	330	55	17%
od	240	62	12%

Total drag at 60 mph, 125 lbs.

SPEEDO ERROR

Indicated	Actual
30	29.4
40	39.0
50	48.1
60	57.0
70	67.0
80	78.6
90	90.0

NOW
it can be told--

THE WORD FROM WARWICKSHIRE

RUMORS OF SOMETHING new from Donald Healey's diggings in Warwick, England, have been filtering through for several months now, and it can finally be reported that our editor and wife paid a personal call on the Healey works last July (en route to Le Mans), where they saw, drove, and photographed the new model.

For all the secrecy surrounding its announcement, the latest Austin-Healey is not startlingly dissimilar to the previous model, but it does have for a powerplant the new B.M.C. engine, a 2639 cc six. The wheelbase has been lengthened by two inches (to 92 in.), and a pair of small jump seats are now provided for what is sometimes called "emergency use"—which for all practical purposes means young children, pets, and packages. The extra length comes in the door panels which are now fitted with exterior handles. Other changes include a new oval grille, a none-too-becoming air-scoop, a fixed (nonadjustable) windshield, and a redesigned folding top. Also, praise be to some clever soul, it is no longer necessary to unload the trunk in order to gas-up, since the filler pipe is now outside.

The official model designation is "Austin-Healey 100 Six," and it is well thus to call attention to the engine because it is a substantial advancement over the former A-90 Four, being much smoother and capable of far higher revolutions. With pushrod overhead valves and a bore and stroke of 3.125 x 3.5 in., it develops 102 bhp at 4600 rpm with an 8.25 to 1 compression ratio. Compared to the Four, the Six's shorter stroke increases theoretical cruising speed in overdrive from 80.1 to 100.4 mph, gear ratios being unchanged.

Austin-Healey fans will find that driving the new car requires no familiarization—everything is the same including steering, controls, and instruments. But the two extra cylinders give the engine an entirely different feel and sound, and performance is deceptive because of the gratifying smoothness. Since facilities were not available for calibrating the speedometer, the editor made a few checks from 0 to 60 mph and spent most of his driving time with the Tapley meter. His impression was that the performance to 70 mph is just about the same as reported for the 100-M model in R & T for March, 1956. That test gave 0 to 60 in 9.6 seconds and the standing ¼ mile in 17.4 seconds. However, the Six revs up to over 5500 rpm, and the factory advises that the acceleration times to speeds of over 80 mph are better than the 100-M. Ultimate maximum speeds in the gears will be 14% higher than the Four, which should help acceleration in the upper speed ranges.

Even so it is apparent that the main intent of the factory is to offer a sports car better suited to a somewhat wider range of buyers than the 100 and 100-M, and the emphasis is on comfort and smoothness rather than sheer performance. Redesigned side-curtains, padded dash, improved weather sealing, etc., all contribute to make the 100 Six acceptable to sports car lovers less hardy than the wind-and-rain-in-your-hair enthusiasts, and two-tone paint, detachable hardtop, overdrive, wire wheels, armrests, etc., are available as more-or-less attractive extras. ●

Behind bucket seats is space for a number of things, all small.

For some reason, size of plastic rear window has been reduced.

With top and curtains up, weather sealing is greatly improved.

Front end changes include new grille and functional airscoop.

In profile the longer lines and smoother top of the Six can be appreciated. Tonneau cover (right) has supporting strut and stays to prevent buffeting underway.

Road Test: AUSTIN-HEALEY 100 SIX

SPORTS car enthusiasts are a strange lot, as most of them (and we) will admit. When a new sports car is offered on the market, they clamor to be the first to have one. But when the first enthusiasm wears off, when the new model becomes more-or-less commonplace (for a sports car), these same enthusiasts will often cast about for something else.

Not so for the Austin-Healey. Announced late in 1952, it was nearly a year before deliveries commenced in any quantity. And for 1953, 1954, and 1955, sales have been consistent and brisk—to the point that supply was even short of demand at times in some areas.

Last spring the B.M.C. organization ceased producing the big four-cylinder "A-90" engine which powered the Austin-Healey sports car. Donald Healey and his corps of engineers at Warwick had been busy developing a side-line business, a 14½ foot "sports boat" of unusual quality. With this project well launched, they returned to sports cars and embarked on "Project 6." The engine for Project 6 was BMC's latest six-cylinder design known as the "A-105" and was well suited to act as a replacement for the old reliable four. The first experimental car completed appears to have been a standard "100" model, with the necessary changes to make room for the longer and slightly heavier six-cylinder unit.

As reported last month, R & T visited Donald Healey last June and was allowed to make extensive tests on the new model. Actually this report is based on experience with two cars, the first experimental machine and a later production version. Out of deference to a tight engine, an all-out high speed run was curtailed, but our estimate is given in the data panel as 105 mph, with a possible variation of 2 mph above or below that figure. The experimental car was definitely "hotter" than the other, and none of the data accumulated on it has been used in this report. However, we might mention that it accelerated from zero to an 60 honest mph in 11.0 seconds, gave Tapley readings in each gear that averaged 11% better than the production model. Both its top speed and peak bhp were given to us as 110.

The basic premise of the new Six is to offer a car which will have a broader appeal and consequently sell in even larger quantities than the Four. Accordingly its main innovations are smoother and quieter running and extra seating space behind the regular bucket seats. In all fairness, it should be mentioned that the two rear "seats" are hardly more than a token gesture and, as far as humans (even children) are concerned, strictly for emergency use. Small and high as they are, however, they may still make the difference for borderline buyers who balk at two front seats only. The actual performance figures are substantially the same as for the 90 bhp model and not as good as those obtained by us last March on the discontinued model 100 M.

Front bucket seats give excellent driving support but rear ones are situated too high and too close for much human use.

Trunk is short on luggage space, but battery has handy master cut-off switch to prevent theft and minimize shock hazard.

New powerplant—B.M.C's "A-105" Six.

Two more cylinders, two more seats, two more w.b. inches, two more mph

	100 SIX	100 FOUR	100 M	100S
Bhp	102	90	110	132
Curb wt.	2480	2150	2385	2115
0-40	6.1	6.2	4.7	4.1
0-60	12.2	11.7	9.6	7.8
0-80	22.5	20.8	17.9	13.9
SS ¼	18.2	18.1	17.4	16.1
Top speed	105	102	109	119
Tested in R&T	1/57	7/54	3/56	9/55

The new Six weighs almost 100 lbs. more than the last of the old Fours, and this of course reduces the acceleration ability by the same ratio (4%). Also, the new Six develops a peak torque of 142 ft-lbs at 2400 rpm, as compared to 150 at 2000 on the Four. This should give the Four Cylinder Club something to cheer over until someone mentions the fact that the Six has a theoretical cruising speed in overdrive of 102 mph as compared to only 80 mph for the 4-banger.

Unlike the 1952 fiasco, the delivery status of the cars will be good, and most dealers will have at least one display car by the end of 1956. The price quoted in the data panel, however, must be considered as tentative, and it may vary somewhat in different localities. The cars will be available in England without some of the traditional Healey equipment, but all cars being imported will have wire wheels, heater, and overdrive as standard equipment just as before.

Externally we find a new grille, an air-scoop, less curved (and fixed) windshield posts, larger doors with exterior handles, outside gas filler, and two-tone paint options.

Continued on page **44**

A-H "expression" is altered by oval grille, airscoop.

R & T ROAD TEST NO. 121

AUSTIN-HEALEY 100 SIX

SPECIFICATIONS

List price	$3195
Wheelbase	92
Tread, f/r	48.8/50
Tire size, in.	5.90-15
Curb weight, lbs.	2480
distribution, %	48.5/51.5
Test weight	2770
Engine	6-cyl., ohv
Bore & stroke	3.125 x 3.50
Displacement, cu in.	161.0
cu cm.	2639
Compression ratio	8.25
Horsepower	102
peaking speed	4600
equivalent mph	110
Torque, ft-lbs.	142
peaking speed	2400
equivalent mph	57
Gear ratios, overall	
O'drive	3.19
4th	4.10
3rd	5.47
2nd	7.84
1st	12.6

CALCULATED DATA

Lbs/hp (test wt.)	27.2
Cu ft./ton mile	84.9
Engine revs/mile	2520
Piston travel, ft./mi.	1470
Mph @ 2500 fpm	102

PERFORMANCE, Mph

Top speed	105
4th (5100)	95
3rd (5500)	76
2nd (5500)	53
1st (5500)	33

see chart for shift points

Mileage range 17/22 mpg.

ACCELERATION, Secs.

0-30 mph	3.8
0-40 mph	6.1
0-50 mph	8.5
0-60 mph	12.2
0-70 mph	16.5
0-80 mph	22.5
standing start ¼ mile	18.2

TAPLEY DATA, Lbs/ton

O'drive	210 @ 56 mph
4th	270 @ 48 mph
3rd	360 @ 40 mph
2nd	470 @ 32 mph
1st	600 @ 21 mph

Total drag at 60 mph, 110 lbs.

SPEEDO ERROR

Indicated	Actual
30 mph	30.0
40 mph	38.8
50 mph	48.4
60 mph	58.3
70 mph	68.2
80 mph	77.6
90 mph	87.7
100 mph	97.8

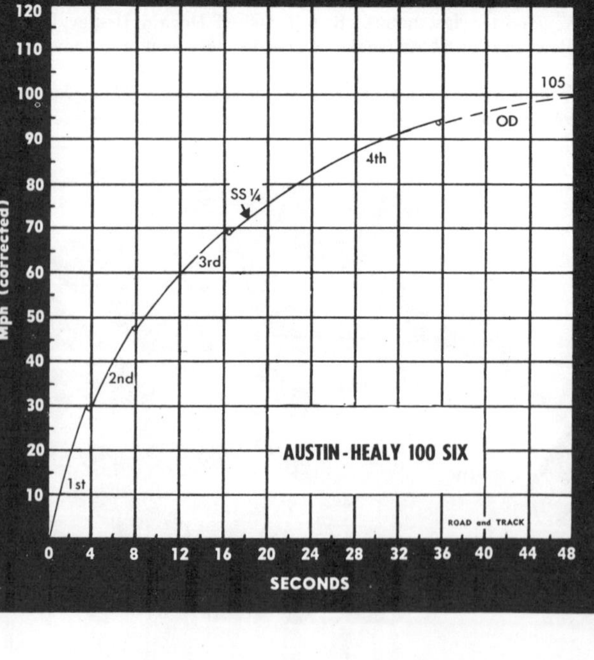

Overall proportions of the Austin-Healey are good. It is one of the few sports cars that does look good with the top either up or down.

ROAD TEST

AUSTIN-HEALEY MILLE MIGLIA

IN A COUNTRY like the U.S., where change is often made just for its own sake, not every import which has remained basically unchanged for six years can continue to sell well. Was the Austin-Healey really so good originally, or have refinements kept it abreast of the times?

The only easily visible changes since the first model have been a new grille and hood (instituted in 1957), outside door handles, and a longer top, necessitated by the addition of the two jump seats. The car is actually 2 inches longer because of the chassis and body lengthening to accommodate the two seats, but this is not readily noticeable. Mechanically, the 6-cylinder Austin A-105 engine was installed in place of the older 4-cylinder A-90 engine, which had been used in the car since its inception.

Now, once again, Donald Healey has brought about some detail changes to keep the A-H up with competition. The new model is the Mille Miglia, named after the race in which the prototype car was first run (and driven to second place in its price class, at an average speed of 75.9 mph by Tommy Wisdom).

The chassis and body of the Mille Miglia are unchanged from the 100-6 (tested in the January 1957 issue, now sold out). All refinements are in the engine, which now delivers 117 brake horsepower at 4750 revolutions per minute. It was formerly rated at 102 bhp at 4600 rpm. The torque has also been increased — from 142 pounds per foot at 2400 rpm to 149 at 3000 rpm. Both power and torque have been raised throughout the speed range, but the greater percentage of improvement is at the upper end.

The chief difference between the old and new engine is in the cylinder head and intake manifold. Separate castings replace the old-style head with its cast-in manifold. Breathing has been much improved by this arrangement, as there are now six intake and six exhaust ports.

The intake ports are round in shape, 1 5/16 inches in diameter, and placed in pairs. Exhaust ports 1 and 6 are rectangular but have the same area as the remaining four, which are round and 1 7/16 inches in diameter.

Two 1¾-inch SU model HD-6 carburetors are attached to the aluminum log-type manifold, which is cast with a partition in the center (between ports 3 and 4). A 7/16-inch-diameter hole is cast in this partition to relieve surging in the manifold and equalize the intake charge, in much the same manner as an outside balance tube.

In order to take advantage of the larger carburetors and better porting, the valves have also been enlarged. The intake valves are now 1.75 inches in diameter, as compared to 1.688 in the previous model. Exhaust valves have been increased in size from 1.420 to 1.562.

Low camera angle gives drama to one of the least pleasing views of the car which, as a whole, is one of the better-looking sports cars on the market. The rear view is handsomer.

Profile of the Austin-Healey is the best view of this well proportioned car.

ROAD TEST AUSTIN-HEALEY MILLE

The compression chamber remains almost the same shape as before but is slightly deeper and has been relieved around the valve seats to aid breathing. The increase in compression ratio — up from 8.25 to 8.50 — was achieved by the use of new pistons with higher crowns.

Dual exhaust manifolds are continued as before. Unlike those on the previous model, they exhaust into twin pipes. Two distinct types of sound emanate from these dual pipes. One is a rather pleasant rap similar to the sound of the XK-120-M, but which at certain rpm gets to be quite annoying to the car's occupants due to the severe "cab noise." It sounds almost as though the exhaust system were welded to the body. A subsequent check revealed that the sound is not bothersome to people outside the car (especially minions of the law) as we suspected it might be.

The other sound coming from the pipes is that of scraping metal whenever any but the most gradual of driveways is entered. It is a continual source of amazement that the designers of the Austin-Healey have apparently made no attempt to increase the ground clearance of the car in general, or at least of the appendages, such as the exhaust pipes, that could be more easily rearranged.

Getting in or out of the Healey with the top up is still a chore. Fortunately, Healey owners spend more time driving the car than they do getting in and out. Entrance and exit are made much easier when the top is lowered; however, lowering and storing the top takes the joint efforts of a score of trained circus roustabouts.

Once one is in, the true bucket seats are fairly comfortable. The controls are well placed with the exception of the choke, which is hidden alongside the steering column and about midway between the instrument panel and the firewall. Healey engineers must feel that the choke is so seldom used that it doesn't need to be located on the panel with the other controls. They appear to be right, too, though California is not the best place to determine this.

The driver's seat is adjustable fore and aft (the passenger's seat is not), and both fold forward to allow admittance to the jump seats from either side of the car. Both front seats are placed in such a manner that the occupants' heads are between the first and second of the three top bows. In spite of this, everyone who rode in the car managed to crack his head against the top bows at least once. Seat belts would be a distinct aid in this respect, although they are not needed for lateral support. The bucket seats are quite effective in holding the driver and passenger in place.

Handling, control, and riding qualities have not been

MM engine is distinguishable by separate intake manifold.

PHOTOGRAPHY: POOLE

The man in command also travels light.

Interior of the MM is unchanged from previous model.

MIGLIA

altered. The ride is on the stiff side, but anyone used to the "sports car ride" will feel right at home. The steering is positive and quick, although a little heavy for so light a car at low speeds, as in town driving.

Visibility with the top down is excellent, as would be expected. It is strongly recommended that a fender-mounted rear view mirror be installed because of the limited visibility with the top up. This, of course, is not a condemnation of the Austin-Healey, but would be most practical for any car with the low seating position and somewhat obscured vision inherent with roadster or convertible tops.

During acceleration tests and top-speed runs the car caused no concern to driver and passenger and seemed to be under complete control at all times. Steering was sure, the suspension coped easily with all road irregularities, and the brakes brought the car to a swift, straight stop.

The clutch and 4-speed transmission worked smoothly, fast starts being made with no difficulty. In making the acceleration runs, the rev limit of 4800 was exceeded only in 1st gear and then only by 400 rpm at the most. It is very likely that acceleration times could be improved by extending the car in each gear. It should also be remembered that the additional passenger's weight adds about a second to the 0-to-60 time.

With the introduction of the MM, the buyer's only choice now is whether to buy the standard or deluxe version.

The standard MM will come equipped with disc wheels; a tonneau cover and heater are the only available extras. The delivered price (plus tax and license) is $2995 on the West Coast. The deluxe model tested comes in delivered form with wire wheels, tonneau cover, heater, direction signals, telescopic steering column and overdrive. The delivered price with everything is $3395 (again on the West Coast). The standard and deluxe models have the same engine. Prices will be about $100 higher in other parts of the country.

Just what is keeping the Austin-Healey in its place in new-car sales? The combination of good looks, compact size, reliability, adequate performance and reasonable price all add up. Some cars have at least two or three of these attributes to a better degree than the Healey, but none of the others has more good points all wrapped up in one attractive package. This, plus the generally excellent service and parts which are readily obtainable over most of the U.S., must be the answer.

ROAD & TRACK ROAD TEST NO. 164

AUSTIN-HEALEY MILLE MIGLIA

SPECIFICATIONS

List price	$3395
Wheelbase	92.0
Tread, f/r	48.8/50.0
Tire size	5.90-15
Curb weight, lb	2600
distribution, %	49/51
Test weight	2885
Engine	6 cyl, ohv
Bore & stroke	3.125 x 3.0
Displacement, cu in.	161.0
cu cm.	2639
Compression ratio	8.50
Horsepower	117
peaking speed	4750
equivalent mph	113
Torque, lb-ft	149
peaking speed	3000
equivalent mph	71.5
Gear ratios, overall	
5th (o/d)	3.19
4th	4.10
3rd	5.47
2nd	7.84
1st	12.6

CALCULATED DATA

Lb/hp (test wt)	24.7
Cu ft/ton mile	81.4
Engine revs/mile	2520
Piston travel, ft/mile	1470
Mph @ 2500 ft/min.	102

PERFORMANCE, Mph.

Top speed (est.)	109.0
best run	104.6
4th (5500)	97
3rd (5500)	76
2nd (5500)	53
1st (5500)	33
Mileage range	17/22 mpg

ACCELERATION, Sec.

0-30 mph	3.3
0-40 mph	5.2
0-50 mph	7.3
0-60 mph	10.4
0-70 mph	14.1
0-80 mph	19.2
0-90 mph	25.6
Standing start ¼ mile	17.4

TAPLEY DATA Lb/ton

5th	230 @ 55 mph
4th	305 @ 50 mph
3rd	400 @ 45 mph
2nd	520 @ 40 mph
Total drag at 60 mph, 102 lb	

SPEEDOMETER ERROR

Indicated	Actual
30 mph	30.4
40 mph	39.1
50 mph	49.0
60 mph	58.6
70 mph	68.4
100 mph	100.0

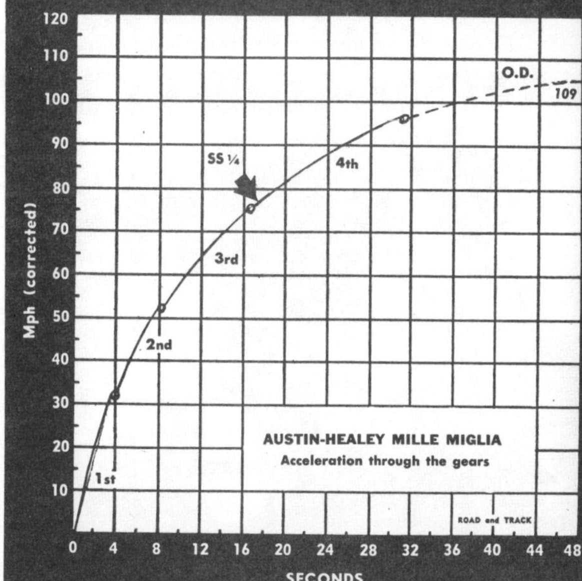

For those with limited budgets, a purposeful, low-cost sports car.

ROAD TEST AUSTIN-HEALEY

THE PARALLELS between General Motors in this country and the British Motors Corporation in England are many. Both are the largest in their respective countries, both build close to half of all cars produced at home. Also, both build sports cars.

But whereas GM builds only the Corvette, the BMC combine has had two, the MG and the Austin-Healey. Now we have a third car to add to the list, the long-awaited Morris Minor version, first predicted in R&T back in November, 1952.

Technically, the new low-priced sports car is to be known as the Austin-Healey Sprite. It is not quite clear why: the car wasn't designed and will not be built by Donald Healey, and the only previous Sprite was a Riley model built before the war. Mechanically, the new Sprite is more Austin than anything else, with engine, gearbox, front suspension and rear axle all being developed from components originally found in the Austin A-30 (now the A-35).

As for appearance, it is obvious that BMC does not follow GM's practice of allowing its artists to relieve their frustrations by styling sports cars. These lines were developed by the body structural engineers to be simple and cheap to form, and to be adequately rigid without the formality of a separate frame. Even the British press says very little about the looks of this new baby—one writer says the shape was developed in a wind tunnel, another opines that the appearance tends to grow on you. We would be inclined to discount both reports. Incidentally, the reason for the frog-like headlights is rather simple: the original design called for concealed lights which popped up when required. Production costs were too high, and the net result has pretty well ruined what otherwise could have had a certain coltish appeal.

Appearance notwithstanding, the Sprite has some very attractive features for the prospective purchaser, among them being 1) low price, 2) comfortable seating and driving position, 3) surprisingly good performance and 4) really excellent handling qualities.

The car stands just 4 feet high with its top up. Entrance or exit is a little awkward, not so much because of the lowness as because the door hinge point is too close to the seat. It is very difficult to get an average-sized foot through the narrow space provided. Once one is seated, the amount of pedal and leg room is a real surprise—more than in the MG-A. The bucket seats are more comfortable than those in the A, but are too nearly vertical for adequate back support on long trips. The low top restricts vision at the sides.

Upraised hood bares entire front end for servicing.

PHOTOGRAPHY: POOLE

Cockpit and instrument panel are starkly functional.

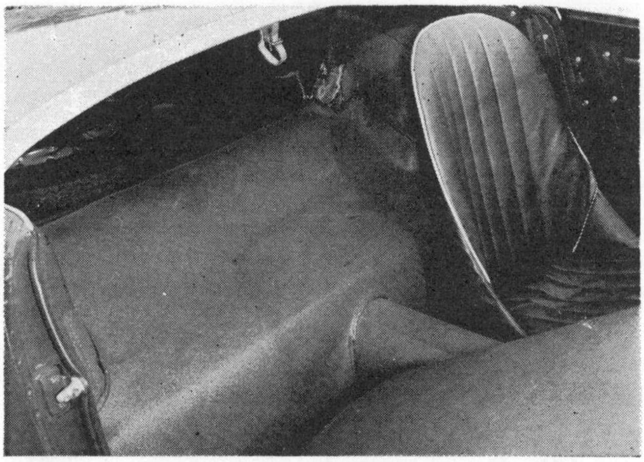

Trunk, spare and tools are reached from behind seats.

SPRITE

The tiny 948–cubic centimeter engine starts instantly and has an authentic sports car rumble that still holds the exhaust note within legal standards. It revs freely and smoothly, with only a little vibration noticeable when decelerating. When it is pressed hard, 6000 revolutions per minute can be attained in the gears without valve lifter noise. We used 5500 rpm as a limit during the acceleration tests because the unit still seemed a little tight, even though the break-in mileage had been reached. Incidentally, the brake horsepower figure quoted in the data panel is the one supplied us, but British reports give quite a different figure of 43 bhp at 5200 rpm. We suspect that 43 bhp is accurate, and that the 48 bhp is under SAE conditions.

The clutch is light, smooth and has a short pedal travel. Its life also may be short, as it definitely slipped after the second all-out standing start test. The gearshift control is beautifully worked out and perfectly located. Unfortunately our test car's lever was so stiff that it could only be moved by brute force from neutral to 1st and from 1st to 2nd. We put over 600 miles on the car ourselves and it did not seem to be "freeing up with use," as claimed. Second gear is much too low except for those drivers who like to use an American 3-speeds-forward pattern. This shows up in the short spurt between dots on our usual acceleration chart. Third and 4th gears shift easily and are well chosen ratios for average use.

For our tastes, high gear is too low. An axle ratio of about 3.9:1 would make 70-mph cruising a little less noisy. The noteworthy silence of 3rd gear would make such a proposal quite feasible and give a brisk city-driving ratio of 5.51:1 overall, instead of the present 5.96 ratio used for 3rd. As it stands now, 4000 rpm in 4th is equivalent to 61 mph, and that's a comfortable cruising speed. Of course 5000 rpm and 76 mph is theoretically safe, but the engine is spinning a little too fast at that rate to feel really serene for hours at a time, even though experience with this powerplant proves that it can be done.

The greatest virtue of the Sprite is without a doubt its excellent steering and handling qualities. With a rack and pinion gear, the turns, lock to lock, are 2.3. This is quick steering, even for a sports car, but once the driver becomes familiar with it there is no objection. In fact, the steering is nearly perfect for the purpose, and light and accurate besides. Cornering characteristics are very close to neutral,

ROAD & TRACK ROAD TEST 176

AUSTIN-HEALEY SPRITE

SPECIFICATIONS

List price	$1795
Curb weight	1460
Test weight	1790
distribution, %	49/51
Dimensions, length	137
width	54
height	48
Wheelbase	80.0
Tread, f and r	45.4/44.8
Tire size	5.20-13
Brake lining area	67.2
Steering, turns	2.3
turning circle	31
Engine type	4 cyl, ohv
Bore & stroke	2.48 x 3.00
Displacement, cu in	57.8
cc	948
Compression ratio	8.30
Bhp @ rpm	48 @ 5000
equivalent mph	78.5
Torque, lb-ft	52 @ 3300
equivalent mph	51.9

GEAR RATIOS

O/d ( ), overall	
4th (1.000)	4.22
3rd (1.412)	5.96
2nd (2.374)	10.0
1st (3.628)	15.3

CALCULATED DATA

Lb/hp (test wt)	36.7
Cu ft/ton mile	71.4
Mph/1000 rpm (4th)	15.7
Engine revs/mile	3820
Piston travel, ft/mile	1910
Rpm @ 2500 ft/min	5000
equivalent mph	78.5
R&T wear index	73.0

PERFORMANCE

Top speed (4th), mph	78.5
best timed run	79.5
3rd (5950)	66
2nd (6000)	40
1st (6000)	26

FUEL CONSUMPTION

Normal range, mpg	29/38

ACCELERATION

0-30 mph, sec	4.7
0-40 mph	8.8
0-50 mph	13.8
0-60 mph	20.8
0-70 mph	35.5
0-80 mph	
0-90 mph	
0-100 mph	
Standing ¼ mile	21.8
speed at end, mph	62

TAPLEY DATA

4th, lb/ton @ mph	190 @ 48
3rd	265 @ 42
2nd	400 @ 30
1st	460 @ 22
Total drag at 60 mph, lb	75

SPEEDOMETER ERROR

30 mph	actual 29.4
40 mph	39.2
50 mph	48.8
60 mph	58.7
70 mph	68.6
80 mph	77.0
90 mph	
100 mph	

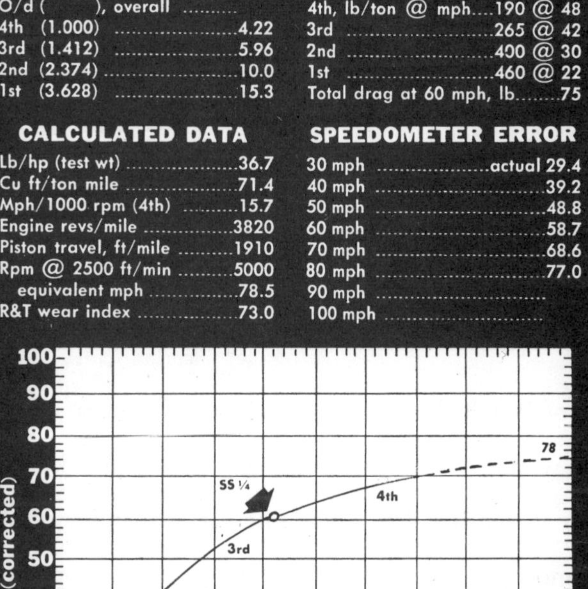

You'll fall head over heels too for this Austin Healey Sprite. A winning beauty—exciting, dashing, quick, responsive. Cleaned up at Sebring. Broke nine International and forty American National Land Speed Records at Wendover, Utah. Top speed, 85 m.p.h. Got everything, but not a bit expensive. The lowest-priced *true* sports car on the market.

Austin Healey sports cars start as low as $1795*

12-month factory warranty on all BMC cars. Literature and overseas delivery information on request.

The AUSTIN HEALEY '3000'. Fabulous successor to the Austin Healey 100-Six which dominated competition in its class. Top speed 115 m.p.h. (2 or 4 seater).

BMC BRITISH MOTOR CORPORATION

Products of **THE BRITISH MOTOR CORPORATION, LTD.,** makers of Austin Healey, Austin, MG, Magnette, Morris and Riley cars. Represented in the United States by HAMBRO AUTOMOTIVE CORPORATION, Dept. 46, 27 W. 57th Street, New York 19, N.Y. Sold and serviced in North America by over 1000 distributors and dealers.

*Suggested retail price at p.o.e.

SPRITE *continued*

Sprite's appearance has been likened to that of Crosley Hotshot, TR-3 and Berkeley. Relocation of headlights and straightening of curved lower hood line (behind front wheels) would be a change for the better.

with just a very small amount of understeer at moderate speeds. It is very hard to force the rear end to break loose, and when it does, the action is neither violent nor hard to control. The front end appeared to be out of alignment on the test car, as the tires were already wearing badly with only 1000 miles on the odometer. This could account for some of the understeer tendency.

At the rear we find a double trailing link type of suspension by quarter-elliptic springs at the bottom and link rods at the top. This system gives favorable low unsprung weight and ties in well with the frameless body construction. Traditionally, quarter-elliptic springs have meant a hard, sometimes choppy ride, and this is true of the Sprite. It does ride exceptionally well for so small a car, and because the independent front suspension system employs coil springs which are softer in action than the rear springs, fore and aft pitching is not too objectionable. In general the ride compares favorably with that standard, the MG, and there is even less roll.

In all fairness there are a few objectionable features about this car, mostly in the nature of bugs which will be corrected in subsequent production models. A marginal clutch has already been mentioned and we accidentally discovered the hand brake lever to be rather shoddy—it broke.

The body panel fit was not too good on this early sample. A check with a 3-foot straightedge revealed some poor forming work, with hollow contours where one would have expected a slight convexity. Raising the hood is inconvenient, the handle and safety catch being so located that your hand scrapes the lower edge of the license plate and the back edge of the front bumper.

The entire front end of the body must be raised, and it is rather heavy, yet flexible. The self-locking struts must be carefully latched one at a time, and then the rather flimsy-looking safety rod can be put in position. Accessibility is surely as good as on any contemporary car.

Because there is no trunk lid, access to the spare tire and tools is awkward, particularly if any luggage is being carried at the time. The seat backs do hinge forward, which makes it possible to get one foot in behind them when loading or unloading the trunk. The omission of a lid is, in this case, more than a cost-saving expedient, for the entire rear end of the body is stiffened considerably by being welded together as one piece. However, the bumper mountings, particularly at the rear, do not look capable of withstanding the American back-till-you-hear-something type of parking. The only objections to the interior were the seat backs, the knuckle-scraping clearance between the shift lever and the instrument panel, and the green direction signal indicator light. Located directly in front of the driver, it is positively blinding at night.

The top covering is removed from the top bows for folding and storage, a feature which certainly saves wear and tear, and the bows nest neatly in rattleproof sockets. However, the simple type of flap which snaps across the top of the windshield has never been noted for keeping out rain, and the fit of the side curtains (or lack of it) also seems to indicate that water leaks may be a problem.

We believe the Sprite has quite a potential for safe sport in production car racing. Undoubtedly we shall soon see the under 1100 cc–class races being overwhelmed with entries under the new marque. A Porsche 1098-cc model could readily defeat the Sprite, but such opponents are rare in this country. The brakes on the Sprite would probably need a little improvement for competition, though they are more than adequate for normal driving.

We found the Austin-Healey Sprite a most significant and exciting new model. It fills the broad gap between the Berkeleys and the MG-A with a genuine sports car at a price. Despite a few obvious faults, it will be popular. It offers more fun per dollar than anything we have driven for a long time.

The Austin Healey '3000'

ROAD TEST AUSTIN-HEALEY 3-LITER

On a jag in a Healey

All of Road & Track's previous road tests of Austin-Healeys—there have been five, not counting this one or the Sprite, since 1954—have been performed in California under practically identical conditions. In the case of the new 3000 model a slight obstacle presented itself. This obstacle was some 2800 miles of road between New York (where the only two 3000's in the country happened to be at the end of May) and California, where the test crew waited with anxious eyes on the deadline for this issue.

Quickly our small but enthusiastic New York staff, consisting of Harvey B. Janes, Eastern Editor, and David E. Davis, Jr., Eastern Advertising Manager, offered a solution. They would drive one of the new cars nonstop from New York to the Road & Track offices in California. Thus we would have our test car in plenty of time and they would be firmly established as heroes, having set all sorts of new coast-to-coast driving records.

They got underway in the early evening on a Monday and, with the aid of a package of innocent-looking but highly potent pep pills, pulled up at the Road & Track

Stock except for its Lucas road lights, the 3-liter looks much better after a bath, even in Los Angeles water.

Here it is as it arrived at the Road & Track office, with a weary Dave Davis and Eastern Editor Harvey Janes.

With the hardtop, a comfortable all-weather town car . . .

And for traveling, better endowed than some sports cars.

offices a little over 57 hours later, full of praise for the car and for their own powers of endurance. When we informed them that a certain French economy sedan had covered the same distance in roughly 54 hours, they were only mildly impressed. The drivers of this French car, they told us, had cheated; they had not stopped to eat along the way. Of the total elapsed time of 57 hours in the Healey, at least three and a half hours had been consumed in various restaurants along the route. In addition, they had been forced to drive through a wild rainstorm and flood in Ohio and had wasted nearly an hour in a bootleg sports car garage just outside of St. Louis trying to replace a silly little rubber grommet that had fallen off the throttle linkage and into a sealed compartment under the instrument panel. In view of all this, our drivers steadfastly and with glassy eyes claimed the following coast-to-coast records: fastest trip in an English car; fastest in an Austin-Healey; fastest in a 3-liter car; fastest in a four-seater roadster with detachable hardtop; fastest by a bearded and mustachioed two-man crew. We might add that it was also the longest distance ever covered in the course of a Road & Track road test.

With the first part of our program successfully completed, we proceeded with the actual performance testing without so much as changing the oil or greasing the car. When the Healey left New York, it had just under 500 miles on the odometer. We had hoped to be able to complete our tests in California without tuning the engine in any way; by the time the car arrived it was so well broken in and running so beautifully—in spite of having been driven hard—that this was entirely feasible.

The speedometer proved to be extremely accurate, which helped our heroes keep up a good on-the-road average. During the acceleration runs it seemed that the gearbox, still somewhat stiff, might be adversely affecting the car's performance; but when we checked the times against those on the 100-6 last year they proved to be just about right in view of the changes that have been incorporated in the new car. The engine of the 3000 is nothing more or less than a bored-out version of the 100-6 powerplant. Displacement is now 2912 cubic centimeters, or roughly 10% larger than before.

On the other hand the rear-axle ratio has been changed from 4.1 to 3.9, which would figure to knock about 5% off the performance all along the line. This leaves a net improvement of 5%. In other words, the increase in engine displacement has more than made up for the change in axle ratio. Acceleration, pulling power on hills, etc., are therefore improved about 5% over last year's car.

On long grades the car pulls beautifully in top overdrive.

The cockpit is the neat and practical one of other models.

PHOTOGRAPHY: POOLE

It looks the same, but it isn't: see text and performance data.

During our top-speed runs we found that if the 3000 is held in 4th gear until about 100 miles per hour, it will fairly leap ahead when overdrive is engaged, and acceleration will continue smoothly and steadily up to better than 112 mph. All of our drivers agreed that in the matter of performance and smoothness the latest Austin-Healey was pleasantly similar to the original XK-120 Jaguar.

Previously, in all our tests on Austin-Healeys, we experienced considerable difficulty in getting the true ultimate top speed. The older models will over-rev in 4th, but when overdrive is engaged at about 100 mph, the car fails to respond. In other words, it takes several miles to obtain terminal velocity. The new car, with the larger engine and closer-ratio overdrive, still has lots of punch at 100 mph and accelerates from that speed to an indicated 120 mph in just over a mile. Tapley readings, for example, are almost exactly 5% better in each gear, as expected, but in overdrive the readings averaged 10% better. The 100-6 used a .778 overdrive ratio with a 4.10 rear axle; the 3000 employs a .820 overdrive ratio which, with the new 3.90 rear axle, gives the same over-all ratio as before in this gear. Here's how the production models compare:

		3000	MM	100-6	100-4
Time,	0-40	5.2	5.2	6.1	6.2
	0-60	9.8	10.4	12.2	11.7
	0-80	17.5	19.2	22.5	20.8
ss ¼		17.1	17.4	18.2	18.1
Top speed		112.5	109.0	105.0	102.0

For the performance tests we adhered to the red line on the tachometer (5200 revolutions per minute). Yet the 6-cylinder engine is very smooth, and it will run well past 5500 rpm with no complaint and no audible sign of valve clatter setting in. The indirect gears are quiet except for 1st, which growls a little. If you get tired of this, it's perfectly possible to forget it and use 2nd gear for starting up. The gearbox was still rather stiff and shifts were hard to make, a feature rather typical of British sports cars for the first few thousand miles. Obviously the gearbox hadn't been used much in the course of the cross-country run.

Handling characteristics are extremely friendly once you get used to the automobile. The steering is quite sensitive to throttle opening—too little produces oversteer, especially on fast bends that have been entered a bit too quickly. Adding more throttle will give fairly constant understeer all the way up to the breakaway point. From there, too much throttle will produce oversteer again, although this can be used to advantage to flick the tail out in tight corners.

Considering the softness of the suspension the ride is surprisingly firm at lower speeds, but above 60 mph it smoothes out considerably and the adhesion seems actually improved, even on rough surfaces. The Dunlop RS-4 tires give excellent traction and on the cross-country trip proved themselves, happily, to be far better on wet roads than previous Dunlops. The 3000 has disc brakes in the front and finned drums in the rear. There is no booster for these brakes, and at speeds below 30 mph the pedal pressure required to stop the car within a comfortable distance is noticeably greater than with the all-drum brake system on the 100-6. At the equivalent of racing speeds the braking is smooth and powerful, with a minimum of nose dive. Even when slight fade is finally produced, after several panic stops from 100 mph, recovery is extremely quick.

The best way to describe the new Austin-Healey 3000 is to say that it is a real enthusiasts' sports car: fun to drive, with lots of performance and good handling and braking characteristics. It could have better cockpit ventilation and seating position, and we still wish that the manufacturer would return to the cleaner styling of the older 4-cylinder cars, but these are only minor grievances. Dollar for dollar this is still one of the top sports cars on the market.

ROAD & TRACK ROAD TEST 211

AUSTIN-HEALEY 3000

SPECIFICATIONS

List price (est)	$3395
Curb weight	2520
Test weight	2820
distribution, %	48/52
Dimensions, length	158
width	60.5
height	49.0
Wheelbase	92.0
Tread, f and r	48.8/50.0
Tire size	5.90-15
Brake lining area	n.a.
Steering, turns	2.5
turning circle, ft	35.5
Engine type	6 cyl, ohv
Bore & stroke	3.28 x 3.50
Displacement, cu in	177.6
cc	2912
Compression ratio	9.00
Bhp @ rpm	137 @ 4750
equivalent mph	112
Torque, lb-ft	175 @ 3000
equivalent mph	71.0

PERFORMANCE

Top speed (5th), mph	112.5
best timed run	115.1
3rd (5450)	81
2nd (5500)	52
1st (5450)	36

FUEL CONSUMPTION

Normal range, mpg	17/22

ACCELERATION

0-30 mph, sec	3.3
0-40 mph	5.2
0-50 mph	7.0
0-60 mph	9.8
0-70 mph	13.2
0-80 mph	17.5
0-90 mph	23.5
0-100 mph	35.0
Standing ¼ mile	17.1
speed at end, mph	79

GEAR RATIOS

O/d (0.82), overall	3.20
4th (1.00)	3.91
3rd (1.31)	5.12
2nd (2.05)	8.02
1st (2.93)	11.5

TAPLEY DATA

5th, lb/ton @ mph	250 @ 62
4th	320 @ 58
3rd	410 @ 48
2nd	540 @ 35
Total drag at 60 mph, lb	105

CALCULATED DATA

Lb/hp (test wt)	20.6
Cu ft/ton mile	92.4
Mph/1000 rpm (5th)	23.7
Engine revs/mile	2535
Piston travel, ft/mile	1480
Rpm @ 2500 ft/min	4290
equivalent mph	101.5
R&T wear index	37.5

SPEEDOMETER ERROR

30 mph	actual 30.2
40 mph	40.0
50 mph	50.0
60 mph	60.0
70 mph	69.2
80 mph	78.5
90 mph	87.7
100 mph	96.8

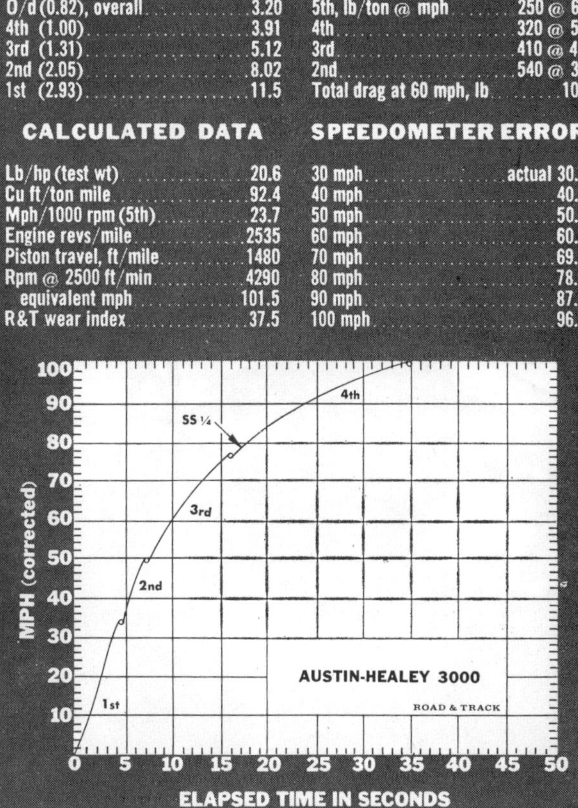

SEBRING SPRITE

With these changes, it's just a little bit better

IN THE TWO YEARS that have passed since the introduction of the Austin-Healey Sprite, the tiny and slightly peculiar-looking British sports car has taken this country by storm. Even though nearly everyone objects to the Sprite's perpetually surprised look, its jaunty behavior and modest price have made it the most popular first-time sports car since the "perpendicular-period" MG-TD was in its heyday.

Although the Sprite and the MG-TD are totally different in design and construction, they have (or perhaps we should say that they are lacking) one thing in common: Neither car has enough power to satisfy some owners. Owner dissatisfaction can, if allowed to run unchecked, bring about serious economic repercussions and no manufacturer wants that. MG offered satisfaction in the form of the Mark II, and BMC—which makes the Sprite—now has available to Sprite owners the equipment used on the "Sebring" Sprites, which have put up such phenomenal times at the circuit for which they were named.

The Sebring Sprite that is the subject of this test comes to us through the courtesy of its owner, Jim Ling, whose use of the car includes both racing and ordinary around-town puttering. Though the Sprite would be run without its muffler in races, we were more interested in the street

Cozy, but quite comfortable.

The front suspension and disc brake.

performance of the car and so conducted our tests with the muffler in place—though this probably reduced the performance fractionally.

Externally, the Sebring Sprite is most noticeable for having knock-off wire wheels and a detachable hardtop, and, if you peer closely enough, you can see that it has disc brakes. These items do nothing for the car's speed but come under the heading of "refinements." To be quite fair we should say that while the disc brakes will not make the car go any quicker in a straight line, they are invaluable in working up good lap times.

The internal modifications, while done very conservatively, are extensive in their scope. The engine is carefully balanced and has special pistons, camshaft, carburetors and manifolding. A special flywheel and clutch take the power to a standard transmission filled with close-ratio gears and then to a numerically higher (4.55 in place of the stock 4.22) rear axle ratio. The suspension is almost unaltered, the only change being a slight modification of the shock dampers.

The changes do not materially affect the feel of the car; it is extremely tractable without a trace of the spluttery idle that so often comes with special camshafts and the like. We were a bit aware of a slight snappishness in the clutch (to which we soon became accustomed) and the marvelous spacing of the gears, which should be made standard on all Sprites.

At the owner's request, and with respect for the number of racing miles that this car has covered, we used only 6000 rpm in our speed trials. For this reason the times given in the data panel are not indicative of the car's absolute best. However, the times are a good indication of how the car would perform in the hands of a non-racing owner, who would be using a similar red line in the interest of reliability.

This test is something of a departure for us, as we do not usually test cars which are not for sale and, needless to say, the supply of genuine Sebring Sprites is limited. We present this as a form of encouragement for Sprite owners, who may have wondered what the penalties of such modifications might be. The truth is that there are none, and you can buy all of the parts, tested and proven, from your BMC dealer, along with complete instructions for bringing your engine up to the Sebring car's stage-5 state of tune.

Here is maximum accessibility.

ROAD & TRACK ROAD TEST 257

SEBRING SPRITE

SPECIFICATIONS

List price	n.a.
Curb weight	1475
Test weight	1870
distribution, %	49/51
Dimensions, length	137
width	54
height	48
Wheelbase	80.0
Tread, f and r	46.4/45.8
Tire size	5.20-13
Brake lining area	n.a.
Steering, turns	2.3
turning circle, ft	31
Engine type	4 cyl, ohv
Bore & stroke	2.48 x 3.00
Displacement, cu in	57.8
cc	948
Compression ratio	9.3
Bhp @ rpm	55 @ 5800
equivalent mph	84.6
Torque, lb-ft. (est)	59 @ 3000
equivalent mph	52.5

GEAR RATIOS

O/d ( ), overall	
4th (1.00)	4.55
3rd (1.35)	6.14
2nd (1.99)	9.05
1st (3.10)	14.1

CALCULATED DATA

Lb/hp (test wt)	37.2
Cu ft/ton mile	73.6
Mph/1000 rpm (4th)	14.6
Engine revs/mile	4120
Piston travel, ft/mile	2060
Rpm @ 2500 ft/min	5000
equivalent mph	73.1
R&T wear index	84.8

PERFORMANCE

Top speed (4th), mph	87.4
best timed run	
3rd (6500)	70
2nd (6550)	48
1st (6600)	31

FUEL CONSUMPTION

Normal range, mpg	25/35

ACCELERATION

0-30 mph, sec	4.7
0-40 mph	7.2
0-50 mph	10.5
0-60 mph	14.2
0-70 mph	20.8
0-80 mph	30.0
0-90 mph	
0-100 mph	
Standing ¼ mile	20.0
speed at end, mph	68.6

TAPLEY DATA

4th, lb/ton @ mph	195 @ 48
3rd	260 @ 40
2nd	370 @ 37
1st	490 @ 30
Total drag at 60 mph, lb	88

SPEEDOMETER ERROR

30 mph	actual 27.4
40 mph	36.5
50 mph	44.8
60 mph	53.3
70 mph	61.8
80 mph	70.7
90 mph	
100 mph	

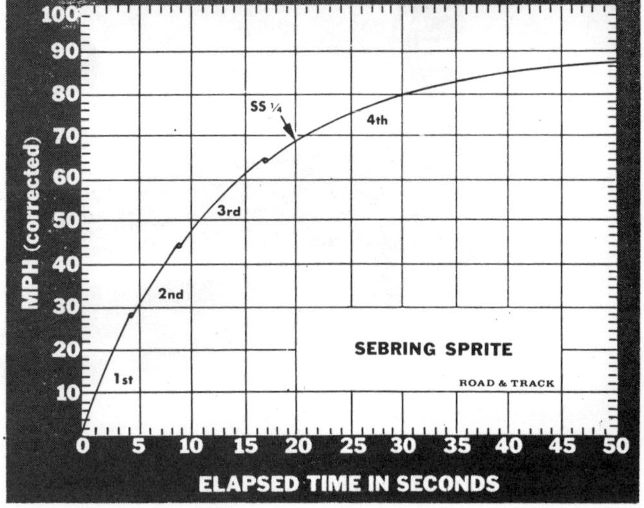

The Pema Sprite Bonnet is made by Peasmarsh Reinforced Plastics, Ltd., Peasmarsh, Nr. Guildford, Surrey, England and utilizes the standard Sprite grille.

Speedwell conversion is made by that well-known speed equipment firm at 763 Finchley Road, London N.W. 11, England.

Ferrari-type hood is made by Dama Plastics, Inc., Swarthmore Industrial Center, Springfield, Penna.

This version of the Sprite hood is made by J. Geddes, 208 17th St., Manhattan Beach, Calif. It uses stock hardware but carries special grille.

FASHION SHOW

unusual "bonnets" for the pampered Sprite

IT IS AN UNWRITTEN RULE that every car owner is just a little dissatisfied with his car in some respect; it may be performance, comfort, economy or looks, but he really isn't happy unless he has something to complain about. We suppose there are people who are completely happy with the car they drive but we've yet to meet any, and if we did meet one, we're afraid he'd probably turn out to be pretty dull.

Sprite owners have their dissatisfactions, too, and to satisfy one of their complaints the enterprising organizations represented on these pages have come up with five new and different front ends for the Sprite. One comes from Switzerland, two from England and two from the U.S.

These replacement noses are all made of fiberglass, but they differ considerably in other respects. Two retain the stock grille, two have special grilles and one has no grille at all. Two have retained the original alligator-type opening method, while the others are hinged at the front. Two have attempted to gain a little performance via improved air penetration characteristics, two are just "different" and one is a vague imitation of a Ferrari Testa Rossa.

Any of these hoods will give individuality to the Sprite, and if we deduct the number of Sprite owners who have no desire to change the appearance of their cars from the grand total of U.S. Sprites, break the remaining group down into the different geographical areas of the country, and then divide by five, we find that there is little chance that anyone who installs one of these adaptations on his car will meet himself coming down the road.

Swiss adaptation made by Emil Frey AG, Flurstrasse 44, Zurich, Switzerland, utilizes stock grille and most of the stock hardware components.

SPRITE II

New look, plus added convenience, equal more desirable car

DIFFERENT AS the Mark II Sprite appears, there are more changes than meet the eye, though the ones you see are the big talking points. From front to rear and from outside in, they are: redesigned front contours with separate hood opening and with headlights in the fenders; redesigned rear contours with a lid for the trunk; enlarged intake valves, longer exhaust valve opening and higher compression to boost SAE horsepower rating from 48 to 50; close-ratio gears, formerly optional, are now standard; rear edge of cockpit has been cut back 12 in. to provide room for a child or some luggage within the cockpit.

The "Two-look" will, we think, go over in a big way. Everywhere we took the test car, it was examined with great enthusiasm and interest. Nearly everyone surmised it was a new model, despite our straight-faced efforts to protect the release date by pretending it was a custom job. Of course, the new model may be accused of some minor loss of personality, but no one can deny that the Sprite II is better looking.

The front fenders no longer lift with the hood, though the curved separation line behind the wheel cutout remains. Headlights are at exactly the 24-in. legal height with the car unloaded. The alligator-type hood is released by an interior pull knob (marked B for bonnet) which should be pushed fully home to ensure that the hood doesn't unlatch accidentally, not forgetting the usual safety catch at the front left corner. The hood is very light and easy to lift . . . quite a change from the heavy awkwardness of the previous model. A long rod is provided to prop it open but you must be careful as you swing it forward not to entangle it with the choke and throttle linkage. Access for maintenance is still excellent, although you'll now need a hoist to grease the front suspension.

The hood extends laterally to the fenders themselves, simplifying the car's lines as well as its production. The fender contours are pleasant ones and they fair nicely with the as-original cowl section and doors. The cheerful grille is gone, replaced by an equivocal aluminum stamping which is better looking in fact than in photo. The large, sensible bumper obscures the attractively complex contour at the bottom leading edge of the fenders.

At the opposite end of the Sprite II, its designers created a new taillight design instead of making do with a standard mass-produced Lucas item. The design they used looks very much like the A-40's, but isn't actually identical. The lenses tidily cap the modest fins, whose height balances the raised line of the front fenders, and also make it easier than ever to reverse a Sprite into a parking niche. To reduce drumming in the fender panel, the rear wheel cutout is flared slightly (for stiffening purposes), which looks odd only because this wasn't done at the front.

The big news at the rear end of the Sprite is the trunk lid. It is lockable, too, and uses a separate key. As at the hood, there is a rod to prop the lid open, though here it is less satisfactory . . . such as when one arm is loaded down.

The spare tire and wheel lie flat on the floor, taking up most of the floor area despite their small size. The 7.2-gallon gas tank is bolted to the bottom of the trunk's flat floor and seems lost in all that space, shared only with the diminutive muffler. With all the retooling for externals, surely it wouldn't have been too much to redesign the floor and the gas tank to maximize the available trunk space. Still, the Mark II is a vast improvement, merely because the access is so much better, and you can pack things in all those small nooks and crannies because you can at least see them now. Now you can get a large two-suiter in (or several duffel bags), along with various small items.

For a car that still has no ashtray, there is the surprising total of five satchels provided for storing items in the trunk. Two are of vinyl plastic, one contains the jack, another the side curtains (a partition separates the pair, giving full protection to the soft plastic windows). A small cloth bag carries the tools, a large one the folded top. A boomerang-shaped leatherette bag, heavily lined with rough felt, carries the tubular top-bows which now split on the car's centerline to ease storage (but require an assistant to install quickly). When all these items are dropped casually into the trunk, one may well wonder where the luggage is going to go. However, there's a nook or a hook for each one and the result is a tidy trunk without a mess of rattles.

As before, the top unfastens completely and if you're neat about it, you can roll it up and store away easily. The tonneau cover, an optional extra that's practically a necessity, is very well designed. It has a bow of its own which uses the same mounting brackets as the top and also splits in two. When it is being used as a boot to cover just the "kiddy cupboard," it passes forward over the bow and then straight down nearly to the floor where it fastens neatly and tautly.

As the trunk is now accessible through a lid, a stiffening bulkhead was put in between it and the cockpit, making up for the reduced rigidity of the body shell. Logically, with the access now at the rear, the trunk was "moved" a foot to the rear and the extension of the cockpit makes space for kids to ride—crosswise, but fairly happily. Legroom here varies from zero to five in., with the accent on the former. A five or six-year-old will even find space enough to curl up for a nap, but on trips of over 100 miles, be prepared for youthful requests to sit in the passenger's lap to get out of the draft.

Cockpit ventilation in the Sprite seems to have just happened. Fully open, it's as fun as can be and the front seat is well protected, apparently by the doors' high sides. But when the top is up, there is a steady blast on the side of your head. With the side screens on, it's very snug and the solitary fault is that the effective heater-ventilator has no cockpit control over temperature. That's done at the rear of the cylinder head, by a miniature water faucet.

Most BMC sports cars have a very stiff gear-change when new; it smooths out with usage but is very awkward for prospects or new owners. It was a pleasant surprise to find that our test car, which had but 600 miles

on it when picked up, could be shifted smoothly and easily. The only fault, also typical of BMC transmissions, was that at a standstill, 1st gear could sometimes be engaged only by previously selecting 2nd. Whether the from-new ease of shifting will be typical of all Sprite IIs is a matter of hope; what is certain is the vast improvement in gear ratios. For a year or so, close-ratio gears as used on the Sebring Sprites have been available, though hard to get. Now they are standard equipment. Here is a table of ratios and maximum speeds for each gear, using 6000 rpm:

GEAR	EARLY SPRITE	SPRITE II
1	3.63—26 mph	3.22—29 mph
2	2.37—39	1.92—48
3	1.41—66	1.36—68
4	1.00—n.a.	1.00—n.a.

We now find that, aside from a rather low 1st, the gears are tightly and evenly grouped and you need never drop below 4200, a certain encouragement to sporting driving and, for road racing, to very hot engine tuning. As the early Sprites could be started from rest in 2nd, we would have been even more pleased if 1st had been raised still more or, happy millennium, if a lower rear axle ratio had been selected to give quieter, lower cruising rpm.

Under the hood, there are further changes, though hardly as dramatic as the hood itself. The compression ratio has been raised to 9.0:1 from 8.3, the SU carburetors have 1¼-in. throats (formerly 1⅛), the intake valves are 1/16-in. larger and the exhaust valve timing has been changed to give 22° more duration by using the same timing as the high-lift, factory option cam. Intake is unchanged. The result is more power without loss of torque. In fact, the torque peak is now at 2750 instead of 3300 rpm, correctly indicating a vast increase in the engine's flexibility, even though all the individual changes normally work in the opposite direction. The number of horsepower gained is small numerically, but significant. Gross or SAE rating changes from 48 to 50 but, more to the point, net, as-installed figures are up from 43 to 46.5. A small price to pay on a car which gets 35 miles per gallon, premium gasoline is now required.

Although there's no accelerator pump to help out, a strong pull on the choke ensures immediate starting, not always the case with SU carburetors. Idle speed varies from 500 to 800 rpm, changing with the nose-up or down attitude.

When starting gently from rest, the clutch engages smoothly, but there is a dreadful rattle from the driveshaft or rear axle. Low gear is very noisy at full throttle, but the other gears are quiet. The ring and pinion had an oscillating noise on the over-run but were silent under power. The exhaust note, with its tempered ferocity, will warm the heart of any enthusiast.

The Sprite is an engagingly simple car, but its mechanical components, all derived directly from BMC sedans, are sturdy rather than sophisticated. When R&T first tested the Sprite three years ago, we criticized on many points. Nearly all have been dealt with successfully. (We aren't sure about the 6¼-in. clutch, although it never slipped during our 700-mile test.) The price in our data panel is that which obtains on the East Coast, and is slightly (about $20) higher in the West. The new list price includes the tach and windshield washers, which were standard "optional" equipment before.

At that, the increase is so slight, as compared with the improvements, that our closing remarks in the previous test seem more pertinent than ever: "Despite a few obvious faults, it will be popular. It offers more fun per dollar than anything we have driven for a long time."

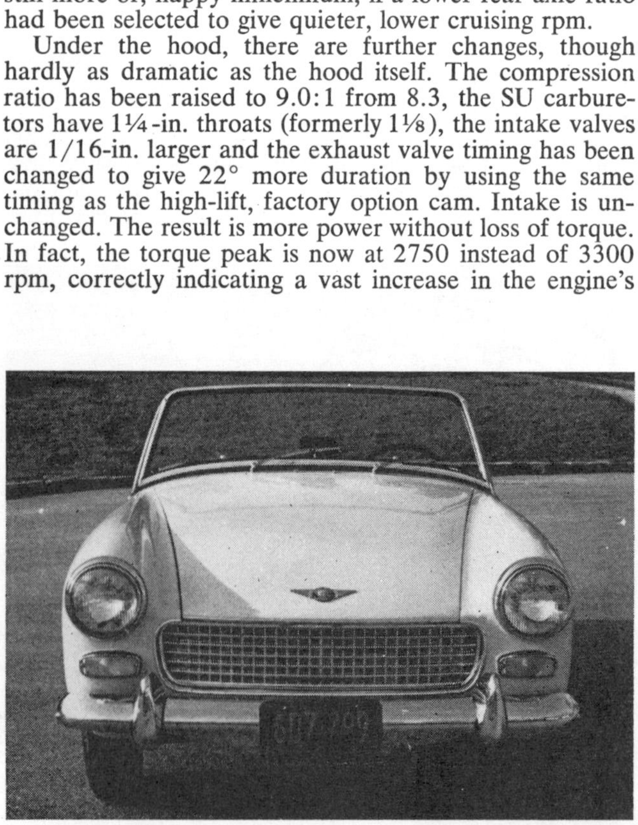

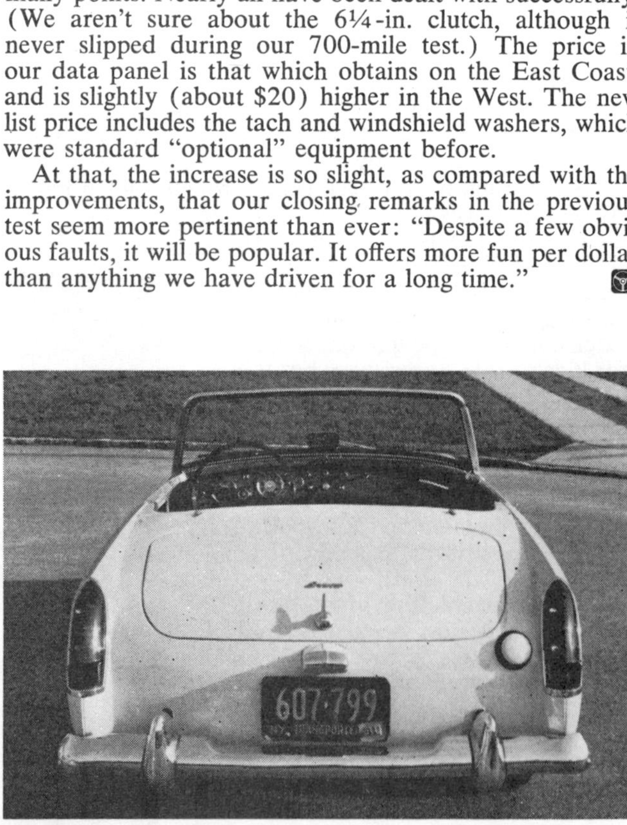

R&T

ROAD TEST

SPRITE II

DIMENSIONS

Wheelbase, in	80.0
Tread, f and r	45.8/44.8
Over-all length, in	136
width	54.0
height	48.8
equivalent vol, cu ft	207
Frontal area, sq ft	14.6
Ground clearance, in	5.0
Steering ratio, o/a	n.a.
turns, lock to lock	2.3
turning circle, ft	31.5
Hip room, front	48
Hip room, rear	n.a.
Pedal to seat back, max.	41
Floor to ground	7.2

CALCULATED DATA

Lb/hp (test wt)	38.0
Cu ft/ton mile	68.7
Mph/1000 rpm (4th)	15.4
Engine revs/mile	3900
Piston travel, ft/mile	1950
Rpm @ 2500 ft/min	5000
equivalent mph	77.0
R&T wear index	76.0

SPECIFICATIONS

List price	$1858
Curb weight, lb	1540
Test weight	1900
distribution, %	51/49
Tire size	5.20-13
Brake swept area	110
Engine type	4 cyl, ohv
Bore & stroke	2.48 x 3.0
Displacement, cc	948
cu in	57.8
Compression ratio	9.00
Bhp @ rpm	50 @ 5500
equivalent mph	84.6
Torque, lb-ft	52.5 @ 2750
equivalent mph	42.3

GEAR RATIOS

4th (1.000)	4.22
3rd (1.357)	5.73
2nd (1.918)	8.09
1st (3.222)	13.6

SPEEDOMETER ERROR

30 mph	actual, 29.6
60 mph	57.8

PERFORMANCE

Top speed (5500), mph	84.6
best timed run	n.a.
3rd (6000)	68
2nd (6000)	48
1st (6050)	29

FUEL CONSUMPTION

Normal range, mpg	32/38

ACCELERATION

0-30 mph, sec	5.4
0-40	8.6
0-50	13.4
0-60	19.6
0-70	31.4
0-80	49.0
0-100	
Standing ¼ mile	21.5
speed at end	62

TAPLEY DATA

4th, lb/ton @ mph	180 @ 45
3rd	250 @ 38
2nd	380 @ 30
Total drag at 60 mph, lb	80

ENGINE SPEED IN GEARS

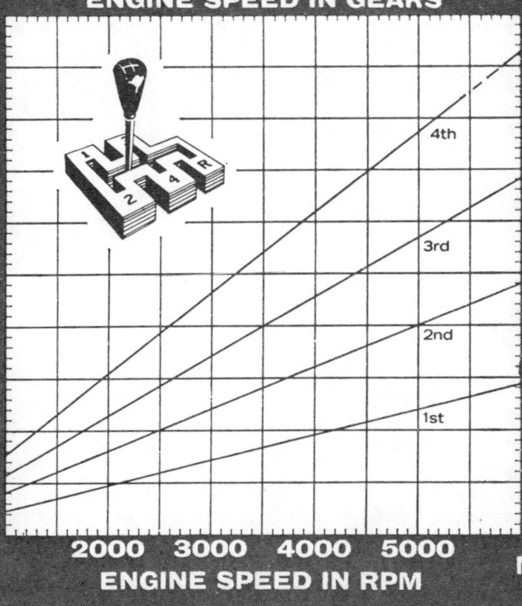

ACCELERATION & COASTING

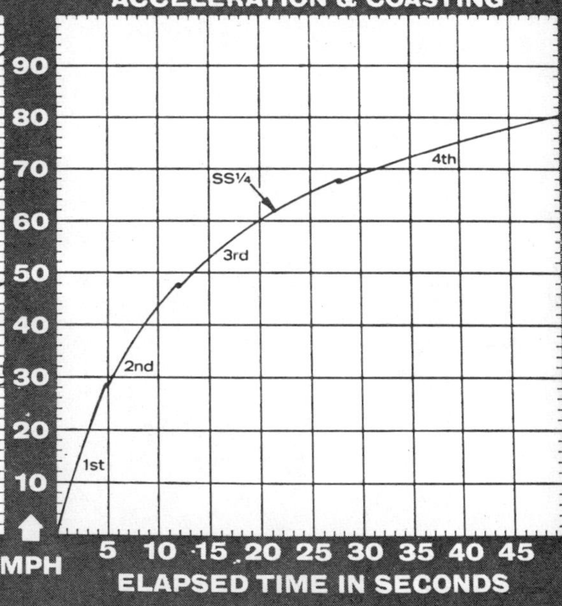

June 1961 Volume 12 Number 10

R&T PS

CHAS. W. HAMILTON PHOTO

Good grief, Charlie Hamilton, a Sprite with a Corvette engine! Bob Knapp, of Palo Alto, Calif., got tired of being shut off at stoplights and decided to do something about it. A Corvette engine (mounted 18 in. further back in the chassis than the Sprite engine had been), even with Powerglide, is a great equalizer. Owner Knapp hasn't had any further trouble with American "iron."

THIS IS THE AUSTIN HEALEY 3000 that didn't win at Sebring. Four makes finished ahead of our No. 33. They cost $2,000 to $8,000 more. Not quite the highest accelerating machine in the world, the Healey will do an honest 120 mph, and 53 Sebring entrants will attest to the fact that you can't beat it by trying to wear it down.

THIS IS THE AUSTIN HEALEY 3000 that could be yours. Some people think it's almost ugly. We feel it is a rather handsome motor car: lean and clean . . . and somewhat mean. All the creature comforts, too—roll up windows, room for a couple of children in back and a true "one-hand" convertible top. All this and 2,380 lbs. of enduring pride.

AUSTIN HEALEY 3000 MK II, A PRODUCT OF THE BRITISH MOTOR CORP., LTD., MAKERS OF MG, AUSTIN HEALEY, SPRITE, MORRIS AND AUSTIN CARS. FOR OVERSEAS DELIVERY INFORMATION, WRITE: BMC, DEPT. Z-7 734 GRAND AVE., RIDGEFIELD, N. J.

AUSTIN-HEALEY 3000

FOR A DECADE, and more, the Austin-Healey sports roadster has been in production, and it has always been a favorite. It was introduced at a time when little choice existed between the extremes represented by the XK-120 Jaguar and the MG-TD, and the Austin-Healey was sold by the boatload here in America. The fact that the car was powered by a truly Stone-Age 4-cyl. engine and had a rather fragile 3-speed transmission did little to inhibit sales; the Austin-Healey was the nicest looking machine available and occupied a spot high on the enthusiast's list of preferred automobiles.

To keep pace with overall advances in automobiles, the lumbering old 4-banger was replaced with a 2.6-liter, in-line 6 (and the car's wheelbase was simultaneously stretched a couple of inches to accommodate the six's length). With the change in engines, and to extend the appeal of the Austin-Healey, the spare tire was relocated and the passenger compartment enlarged to include "occasional" seating behind the driver and passenger seats. And the old 3-speed transmission was exchanged for a more sturdy and versatile 4-speed unit.

Subsequent upgradings have seen the engine further enlarged (to just under 3-liters) and a host of detail improvements; none of which really changed the character of the car very much. Both 2- and 3-carburetor versions (from two 1¾-in. to three 1½-in. SU carburetors in 1961) of the "3000" have been produced and the car's performance has been steadily improved, along with the creature comforts.

This year, the greatest improvement of all has been made: a change to roll-up windows, the *sine qua non* of the modern sports car. Getting the sliding glass and the cranking mechanism into those thin roadster doors must have been a pretty good trick, but it has been done. The cranking system is fairly direct acting, and the new windows zing up and down in a flash, which is certainly an improvement over the old side-curtains—and there are small, pivoting wind-wings that can be opened for draft-free ventilation. To those of us who remember the infamous "side-flap" model of the early days, these concessions to easy-arm-signals and fresh air are enough to bring tears of relief.

We did, however, discover that (probably due to the scarcity of space within the doors) the side windows were rather poorly braced. When cranked uo, they would flutter and rattle, and the frame around the wind-wing—which had a channel to guide and steady the window—was not strong enough to stop the shaking. Only when the top was up and the windows cranked all the way up into the top frames, did the tremors cease.

The top itself is new, too. Unlike the earlier efforts, the top that goes with the roll-up windows is a true convertible, as opposed to roadster, top. It attaches securely to the bodywork around the back, and has linkages that really do guide it into place. No guessing games or step-by-step instructions or anything; just grab the convenient handle on the windshield bar and haul it up into place. A pair of simple, over-center latches are provided to dog it down. When up, it is air-tight, water-tight and (here is a novelty) it does not drum much. Added to these virtues is a big back window, which on warm days may be dropped down (by releasing a couple of catches) to give yet more fresh air.

Unfortunately, here again someone suffered a lapse, or something. When the time comes to drop the top back down into its well (where it can be stowed neatly and well out of sight) there is a serious snag. If the top is just tossed back—as it very easily can be—the folding links will scissor ragged little holes in the top material and the owner will say something heartfelt and forceful. Naturally, this top-tearing does not occur every time the top goes down, but it will happen about every third or fourth time, and that is certainly often enough to make an owner pretty grouchy. So, do be careful, and make certain that the fabric is held clear of the framing as you lower the top.

Of lesser importance than the top and side windows is the new windshield. Previous Austin-Healey windshields were nearly flat; the new one (still a simple curve) wraps around and fairs into the window posts a lot more smoothly. Better vision and less wind turbulence are claimed. We are not certain about these factors, although the vision was excellent, as was the case in the older model, and it does *look* more streamlined. Certainly, the new windshield is more modern and attractive.

Mechanically, the latest version of the Austin-Healey is very little changed over last year's car. The engine is essentially the same, and produces the same power, but the 3-carburetor setup (1½-in. SUs) has been abandoned and the engineers have reverted to 2 (1¾-in. SU) carburetors. This would seem, on the face of it, to be a regression, but the car's performance does not seem to have been reduced by the change. Indeed, in comparing reports on the 3-carburetor car with the figures we obtained with this new machine, we found evidence that the engine may be slightly more powerful with one less carburetor.

The transmission has been slightly re-designed—given a new shift mechanism that is more positive and smoother than before. Externally, this change manifests itself in the shift lever, which was rather long and snaky, but is now straight and grows right up out of the transmission tunnel next to the driver's knee. Unfortunately, the re-design was not carried far enough to eliminate the pronounced gear-whine, and there is still no synchromesh on first gear; a curious deficiency considering the car's price.

The rear suspension, an ordinary live axle carried on leaf springs, has not been changed, but the A-arm type front suspension has been given slightly stiffer coil springs to improve the car's roadability. That is, at least, the purpose of the change in spring rate, but we are not convinced that it has actually improved things. The Austin-Healey has never been the fastest-cornering car in this world, but it always had great stability. Now, (and we checked this on two different examples of the new car) the car is just a trifle twitchy when pushed hard, and it is persuaded to travel in a straight line only with some difficulty. Even on a smooth road, and in relatively still air, the Austin-Healey would wander from side to side, and the driver is forced to make endless small corrections to hold his course.

Interior layout and appointments in the Austin-Healey are good—but with a couple of minor points of irritation. The seats follow the classic bucket pattern closely, with a contoured cushion on which one sits and a seat back that wraps well around at the sides to give lateral support. These seats were comfortable, but would have been better if they had been raked back more. As they were (and there was no adjustment for rake) one is forced to sit very upright, which may be good for the posture but is not entirely comfortable. The back seats, such as they are, cannot be seriously intended for occupancy by anyone but children, and as we were unable to get any children to render an opinion, we cannot report on that part of the seating package.

Trunk space in this car is all but non-existent—unless you happen to be filling it with sand, or some other substance that will flow into all kinds of irregular corners. The spare tire, which was crowded down to the trunk floor when the Austin-Healey was converted into a 2/4-seater, effectively prevents the

The unexciting but exceedingly reliable Austin "six."

The Austin-Healey was, and is, a nice-looking car.

AUSTIN-HEALEY 3000

stowage of hard luggage; a soft duffle bag could be accommodated, but that is about all.

Serviceability is a mixture of good and bad: all of the points requiring attention are easily reached, but if the driver's handbook instructions are followed, the points are going to be often reached, too. For example, the manual calls for attention to no less than 25 lubrication points every 1000 miles. With the passing of 3000 miles, the ritual becomes even more involved, and at 12,000 miles there is a check list to be followed that would keep a corps of mechanics busy for a fortnight. They will be even busier at 24,000 miles, when the manual politely requests that, in addition to everything else, the sump be removed and cleaned. This kind of hypochondria contrasts rather oddly with the modern trend toward sealed, self-lubricating bearings and minimal service. On the other hand, the Austin-Healey has demonstrated a rather impressive reliability over the years; whether this is because the owners follow the service procedures, or in spite of the fact that they ignore them, we are not prepared to say.

As a touring car, the Austin-Healey has a great many virtues. Its big 6-cylinder engine does not have to work very hard to propel the car down the road, and the overall drive ratio provided by running in overdrive gives a relatively low engine speed even at 100 mph (4200 rpm, to be exact).

The new top offers good weather protection, visibility.

We would have strongly preferred that there not be quite so much mechanical noise at high speeds, and other cars have proved that a harsh ride like that of the Austin-Healey is not really essential to good road manners, but apart from those objections it was all right. The only thing that bothers us is that the Austin-Healey is going to have to sit on the same showroom floor with BMC's MG-B, and it is in all respects save performance substantially inferior to that new model. Perhaps it is time for an Austin-Healey "B."

Instruments are clustered and can be easily read.

An old-fashioned prop holds open the trunk lid.

ROAD TEST

AUSTIN-HEALEY 3000

DIMENSIONS

Wheelbase, in. 92.0
Tread, f and r 48.7/50.0
Over-all length, in. 157.5
width 60.5
height 50.0
equivalent vol, cu ft. 274
Frontal area, sq ft. 16.8
Ground clearance, in. 4.5
Steering ratio, o/a. . . . . . . n.a.
turns, lock to lock. 2.5
turning circle, ft. 35
Hip room, front. . . . . . 2 x 18.0
Hip room, rear. 36.0
Pedal to seat back. 43.0
Floor to ground. 9.0

CALCULATED DATA

Lb/hp (test wt). 20.3
Cu ft/ton mile. 94.5
Mph/1000 rpm (o/d). 23.7
Engine revs/mile. 2535
Piston travel, ft/mile. . . . 1480
Rpm @ 2500 ft/min. 4290
equivalent mph. 101
R&T wear index. 37.5

SPECIFICATIONS

List price. . . . . n.a.
Curb weight, lb. 2430
Test weight. 2760
distribution, % 48/52
Tire size. 5.90-15
Brake swept area. . . . . n.a.
Engine type. . . . . 6-cyl, ohv
Bore & stroke. . . . . 3.28 x 3.50
Displacement, cc. 2912
cu in. 177.7
Compression ratio. 9.03
Bhp @ rpm. . . . . . 136 @ 4750
equivalent mph. 112
Torque, lb-ft. . . . . 167 @ 3000
equivalent mph 71

GEAR RATIOS

o/d (0.82). 3.20
4th (1.00). 3.91
3rd (1.31). 5.12
2nd (2.05). 8.02
1st (2.93). 11.5

SPEEDOMETER ERROR

30 mph. . . . . actual, 29.2
60 mph. 59.4

PERFORMANCE

Top speed (4850), mph. . . . 115
best timed run. . . . . n.a.
3rd (5200). 77
2nd (5200). 49
1st (5200). 34

FUEL CONSUMPTION

Normal range, mpg. 15/21

ACCELERATION

0-30 mph, sec. 3.5
0-40. 5.5
0-50. 8.2
0-60. . . . . 11.2
0-70. . . . . 14.7
0-80. . . . . 19.8
0-100. . . . . 37.0
Standing ¼ mile. 17.6
speed at end. 77

TAPLEY DATA

4th, lb/ton @ mph. . 265 @ 54
3rd. . . . . 365 @ 46
2nd. . . . . 540 @ 41
Total drag at 60 mph, lb. . . 120

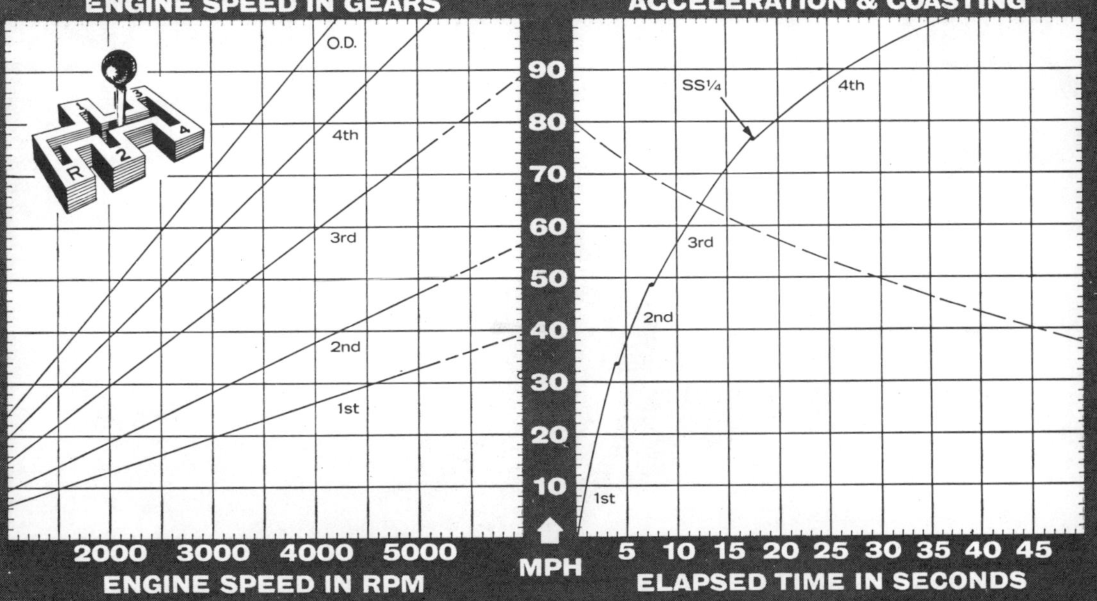

AUSTIN HEALY 100-SIX

CONTINUED FROM PAGE 18

The top is, of course, larger and, we think, more graceful than before. However, it proved to be quite a chore to stow the top properly so as to fit behind the removable rear seat back—without an instruction book. New aluminum-framed side-curtains are supplied which fit very neatly and incorporate a sliding plexiglass panel. Luggage space is reduced considerably by the "jump" seats and the new battery location (see photo).

A first impression on driving the car is that the steering wheel, though adjustable, is too close to the chest and too high, but this soon becomes less noticeable. The engine is very quiet at idle, but has just the right amount of exhaust noise when underway to satisfy most owners. The four-speed gearbox is the same heavy duty unit supplied on the 100 M and 100 S models, first gear being unsynchronized and quite low. This gear gives a tremendous burst to just over 30 mph, but second gear starts are perfectly feasible and can be accomplished without excessive clutch slip. The shift is a little stiff, but the synchro-cones work perfectly, and you can literally jam the lever into place with no clashing. Unfortunately there is still some heat around the feet and legs which seems to emanate from the tunnel/dash area and a side cowl ventilator would be a good item for some accessory manufacturer.

The steering is unchanged and as before is excellent. It requires only 2.8 turns from lock to lock, and there is just enough understeer to avoid backing off on the wheel in a fast corner. The highly satisfactory brakes are also unchanged, but for the competition minded, Girling disc brakes will be available later.

It is always difficult to evaluate the riding qualities of a car, but our impression is that this car rides somewhat more firmly than the superseded model 100, and possibly about the same as the competition-sprung 100 M. This was with the tires at 30 psi, and for a car designed to appeal to enthusiasts with a family we feel that the original, softer A-H ride would be more acceptable to more people. One incidental advantage of this suspension is that at no time did we encounter rear spring bottoming during fast starts from a standstill, which we reported in our original 1954 A-H test.

What is being done for the enthusiast who wants to run occasionally in the production sports car class? There is no definite answer to this question, but the Bonneville car which ran last summer had an experimental cylinder head which may ultimately become a "Le Mans Kit." Certainly the drop in stroke from 4.375" to 3.50" along with other basic design details will make a more rugged powerplant for the A-H. Something like 140 to 150 bhp should be possible at 5000 rpm without sacrificing reliability.

In summation perhaps we should admit that we like the "looks" of the old car better than the new, but this Six has a real potential both in immediate sales and in future development towards even more performance. ●

PS

Happiness is your own parking place.

PHOTO BY ED WHITE

AUSTIN-HEALEY SPRITE 1100

More under the bonnet, and better to boot

UNLIKE THE U.S. AUTOMOBILE INDUSTRY, most overseas manufacturers produce any given model for several years, with only minor detail changes made each year. Such is the case with British Motors Corporation's Austin-Healey Sprite, although in this case the changes were more than *minor* details and constitute a genuine upgrading of the car.

Briefly, the changes were: An increase in engine displacement from 948 to 1098 cc, which raised the horsepower rating from 46.5 to 55 bhp (still at 5500 rpm); the clutch diameter was increased one inch; baulk ring synchromesh (top 3 gears only) for the transmission; a change in front brakes from drums to discs; new seats which are both reshaped and more thickly padded and the addition of carpeting in the interior to replace the "unborn Gila Monster hide" that came with the original Sprites.

The additional engine displacement was brought about by an increase in bore and a new, longer stroke crankshaft very similar to the ones used in the MG-1100 sedans. In fact, the upper part of the Sprite engine is now almost identical to the MG-1100.

The interior trim on our test car was well detailed and everything seemed to fit, and work, properly. The seat design, in addition to being better looking than before, is also more comfortable and the seats offer good lateral support. The carpeting serves the double function of helping the looks and keeping out some of the road noise. It was the consensus of this office that the Sprite now looks more like a car than a toy—aided by more attention to assembly detail than has been previously shown.

Instrumentation consists of a large-diameter tachometer—red lined at 6000, with an orange warning section from 5500 to 6000 rpm—and matching speedometer set directly in front of the driver; a fuel gauge at the left and a combined oil pressure/water temperature gauge at the right.

In view of current automobile design practice it is surprising that BMC has continued the separate key and starter actuating (most cars now use key-start ignition switches), and even more surprising that it is a pull cable rather than a push button. In further contradiction to the trends, the heater control is a knob similar to the starter knob—but pulls to shut *off,* pushes to bring up cockpit heat.

The steering wheel is set directly in front of the driver (not all the smaller cars have them so placed because of lateral space problems) but the column angles off slightly to the right. This isn't too noticeable and didn't cause any bother. Foot pedals are placed in the spots one would expect to find them and about as close together as possible and still leave room for the large American foot (we've come to the conclusion that British and Italian feet must be smaller than their American counterparts).

All hand controls fall readily to hand—and if one isn't careful, the hand on the shift knob can become battered and bruised from hitting the lower part of the panel on shifts to third. The lower edge of this panel is padded along its full length which prevents permanent damage to knuckles (and passenger's knees) from accidental contact with the edge.

There are still no door locks or outside door handles—it is necessary to reach inside for the handle, which means sliding forward the split side window if the top is up. These sliding perspex side windows, mounted in metal frames, are removable and are far superior to side curtains for both noise and weather protection, but still not as good as roll up windows.

The absence of window winding mechanism in the door does leave room for a hollowed-out area, a necessity if there is to be elbow room in a tiny car, and also helps a great deal in keeping the overall cost of the car down. The side window frames on our test car would not stay in place well because the attaching bolts were too short, so most of the testing was done with the side windows removed. Due to the Sprite's top and side window design, visibility is good even with frames in place and top up—a feature not all convertibles can claim.

Aside from the increased comfort of the seats and the lower noise level, the additional displacement of the engine has added the most pleasure to driving the Sprite because of the extra torque available. Even though 8.5 ft/lb doesn't seem like much increase, it represents 16.2%.

Our top speed runs netted only a fraction more than the 948-cc Sprite II but we believe that a well-broken-in Sprite 1100 should be able to add a few mph to our top speed figures. Even more important to the driver though is the top gear hill climbing and easier passing ability offered by the better torque characteristics throughout the entire range.

The disc brakes in front have contributed a great deal to the safety margin because the former small drums were somewhat marginal when the car was being driven near the maximum. Handling has been unaffected by any of the changes and the Sprite is still one of the most responsive cars on the road.

The steering, rack and pinion, is quick and accurate, although a small amount of road shock is transmitted back to the steering wheel. The Sprite can be driven through a series of fast S bends with only the slightest movement of the wheel and although the tail tends to drift out, from body roll, on entering a curve, it stabilizes almost immediately and gives the feeling of being a completely neutral steering car.

Due to its responsiveness and neutral handling it is an excellent car for the beginning sports car driver; and with the nominal initial cost and good fuel economy it is also ideal for the impecunious.

The ride is severe by comparison with most cars and def-

SPRITE 1100

initely on the stiffish side. Yet, with two persons aboard, a slight dip or thank-you-ma'm would cause the car to bottom at the rear. The good dampers brought the car back to level with a minimum of vertical oscillation, however.

It was our opinion, at first, that the fuel capacity (7.2 gal) is too limiting, for even with its low fuel consumption the Sprite can safely go just over 200 miles between gas stops. Looking back on our experience in the car, we're not sure we'd want to go more than 200 miles between stops anyway because in spite of the comfortable seats, we found the combination of wind noise and relatively harsh ride tiring.

Luggage space remains unchanged from the Sprite II and its alter ego the MG Midget. Each door contains a map pocket which would do for the usual detritus found in a glove compartment (which the Sprite lacks) with the advantage/disadvantage that it would be difficult for either driver or passenger to reach the compartment on the opposite side. There is room behind the seats for a pretty fair amount of luggage and we've been told that a third person can even ride back there, seated crosswise in the car, but we find this hard to believe and certainly don't recommend it.

The only lockable space on the car is the trunk compartment which holds the spare tire, and top and side curtains when down, and has room for a fairly good sized suitcase.

We have found that Sprite owners are generally unanimous in their praise of the car for daily use and we know several who have put on a large number of miles with very little or no expense above the usual gas, oil and periodic servicing. The biggest and most often heard complaints stem from the annoyance of things coming loose, usually around the cockpit, and from the seemingly incurable oil leaks in the engine.

Be that as it may, we've said it before, and we have to say it again: The Sprite (and Midget) offers more fun per dollar than any car we can name—providing the car is accepted for what it is.

R&T

ROAD TEST

A-H SPRITE 1100

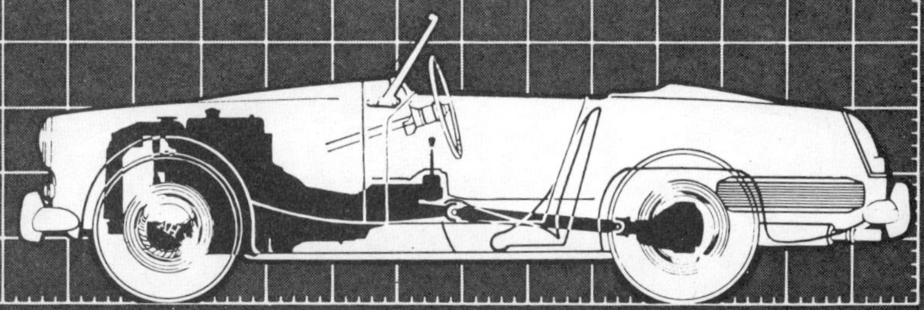

DIMENSIONS

Wheelbase, in	80.5
Tread, f and r	47.2/45.0
Over-all length, in	138
width	54.0
height	47.8
equivalent vol, cu ft	206
Frontal area, sq ft	14.3
Ground clearance, in	6.2
Steering ratio, o/a	n s.
turns, lock to lock	2.3
turning circle, ft	30
Hip room, front	2 x 16
Hip room, rear	n. a.
Pedal to seat back, max	41.0
Floor to ground	7.2

CALCULATED DATA

Lb/hp (test wt)	34.2
Cu ft/ton mile	80.4
Mph/1000 rpm (4th)	15.4
Engine revs/mile	3900
Piston travel, ft/mile	2140
Rpm @ 2500 ft/min	4550
equivalent mph	70.0
R&T wear index	83.5

SPECIFICATIONS

List price	$1985
Curb weight, lb	1560
Test weight	1880
distribution, %	51/49
Tire size	5.20-13
Brake swept area	190
Engine type	4-cyl, ohv
Bore & stroke	2.54 x 3.30
Displacement, cc	1098
cu in	66.98
Compression ratio	8.90
Bhp @ rpm	55 @ 5500
equivalent mph	84.6
Torque, lb-ft	61 @ 2500
equivalent mph	38.4

GEAR RATIOS

4th (1.000)	4.22
3rd (1.357)	5.73
2nd (1.918)	8.09
1st (3.222)	13.6

SPEEDOMETER ERROR

30 mph	actual, 29.3
60 mph	57.8

PERFORMANCE

Top speed (5530), mph	85
Shifts, rpm-mph	
3rd (6000)	68
2nd (6000)	48
1st (6050)	29

FUEL CONSUMPTION

Normal range, mpg	30/36

ACCELERATION

0-30 mph, sec	5.4
0-40	8.3
0-50	12.5
0-60	18.3
0-70	27.8
0-80	42.5
0-100	
Standing ¼ mile	20.9
speed at end	63

TAPLEY DATA

4th, maximum gradient, %	9.8
3rd	13.5
2nd	19.3
Total drag at 60 mph, lb	75

ENGINE SPEED IN GEARS

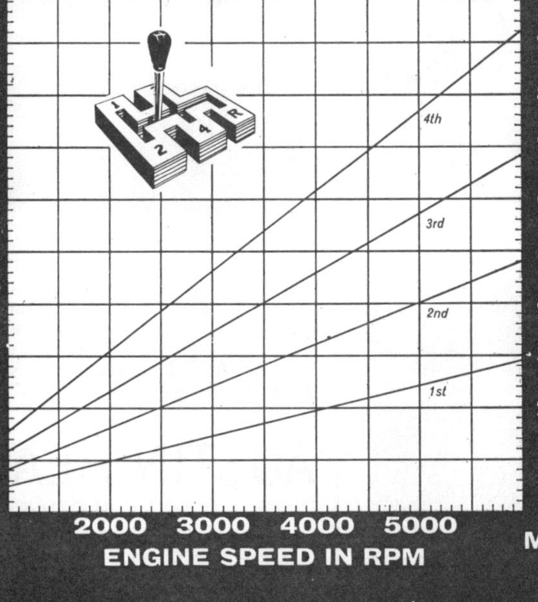

ACCELERATION & COASTING

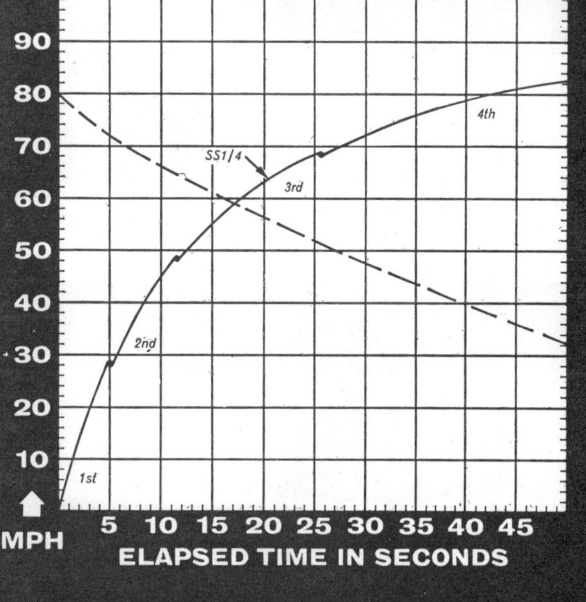

HERE ARE A FEW REASONS WHY THE NEW AUSTIN HEALEY 3000 WILL INTRIGUE YOU

Patrician breeding shows in every sculptured line of the new Austin Healey 3000. She's dressed for the Sixties in a wrap-around windshield, draft-free vents, and a multitude of other refinements for your pleasure and comfort. Her new vinyl-

treated fabric hood can be raised or lowered with one hand. She's graduated from a roadster to a true convertible. In

the sunlight or under the stars, her high-stepping performance lives up to her looks. The 6-cylinder overhead-valve mill will carry you at sustained high speeds...and she's a real neck-snapper if you accelerate at 70. Big disc brakes mean extra safety, and for the statistically-minded, the Austin Healey 3000 has a compresslon ratio of 9:03 to 1, with a 3.5-in. stroke in a 3.282-in. bore. The smart efficiency of the cockpit will gladden your heart. The cowled dials are grouped for quick reading; shift and

toggle switches are handy. Functional luxury—for a commute or a lazy country ramble. Owners are often prone—to ad-

mire the Austin Healey 3000 from every angle. Each subtly sculptured line flows into the next with aerodynamic logic; the

Q.E.D. is a car of impressive appearance and eager power. You can take the tads

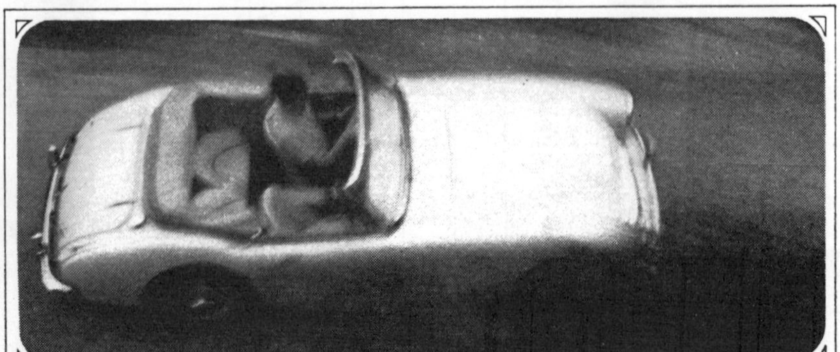

along now and then, because this car is an "Occasional four". But before a long trip, recite three times: "Two's company. . . ." You'll get a renewed sense of pleasure every time you step into the

cockpit of this responsive little beauty. Do it soon, because nothing will be so convincing as your own sense of power and pleasure at the wheel.

Rambunctious tomboy sister of the 3000 is the Sprite. She's a classy little package whose twin-carb power will take you over 220 flashing miles on one filling of the 7¼-gallon tank. Stowable hood, oversized brakes, an endurance winner at Sebring. At P.O.E./N.Y., $1875, you can afford to flirt a little, can't you?

AUSTIN HEALEY

3000 MK. II CONVERTIBLE

PRODUCT OF THE BRITISH MOTOR CORPORATION, LTD., MAKERS OF MG, AUSTIN HEALEY, SPRITE, MORRIS AND AUSTIN CARS

FOR OVERSEAS DELIVERY INFORMATION, WRITE: BMC, DEPT. Z-3, 734 GRAND AVENUE, RIDGEFIELD, N. J.

INNOCENTI S

Combining Italian coachwork with the familiar A-H Sprite mechanical bits

A NEWCOMER AMONG the sports cars currently offered on the American market is the Innocenti S, which, on closer inspection, turns out to be our old friend the Austin-Healey Sprite (or MG Midget) neatly disguised in Italian bodywork.

Innocenti (pronounced "Innochenty") is better known as the manufacturer of the Lambretta motor scooter, an extremely successful Italian 2-wheeler. In order to expand its activities, Innocenti reached an agreement with the British Motor Corporation to assemble and market certain BMC products in Italy, and among them is the Sprite. This would appear to be an excellent arrangement for all concerned because it gives BMC an outlet in a Common Market country, and it permits Innocenti to offer a range of small cars suitable for the Italian market without investing in extensive manufacturing facilities.

Although Italy is thought of as a car manufacturing country of some stature, Fiat is the only producer of small sedans and, while the range and quality of its products are well known, there is a definite market for other makes of automobiles more utilitarian than Lancias and Alfas. This accounts in part for the multitude of Fiat derivatives which abound in Italy and offer something more and different than the production Fiat. By offering its own version of the BMC products, Innocenti has successfully tapped a part of the market and is now well established.

Another consideration is that there has been a surprising increase in and re-distribution of wealth in Italy since the war and a large car-owning middle class is rapidly emerging. The Innocenti range is obviously aimed at this new class and, having previously sold them Lambretta scooters, the company is now in a good position to trade them into small cars.

One of the most striking differences between the Innocenti S and the Sprite or Midget is the price; at $2920 West Coast POE, the Innocenti is some $800 more than its English cousins with similar equipment. Undoubtedly, a part of the price difference results from the relative inefficiency of a method of automobile construction which involves shipping components from one country to another and then assembling the automobile from parts from a variety of sources. However, this does not add to the value of the finished product.

Surprisingly enough, the car is almost entirely BMC, with the exception of the body, and such components as the electrical system, the instruments and some other more insignificant items. On our test car, the Veglia instruments

INNOCENTI S

AT A GLANCE...

Price as tested	$2937
Engine	4 cyl, ohv, 1098 cc, 58 hp
Curb weight, lb	1560
Top speed, mph	85
Acceleration, 0–60 mph, sec	18.4
Passing test, 50–70 mph, sec	13.9
Overall fuel consumption, mpg	31.1

INNOCENTI S

were all strictly Italian, with a speedometer reading in kilometers and gauges for *acqua, olio, benzina* and *giri*. However, this presents no problem, with the exception of the speedometer, which requires a quick calculation to find out whether one is speeding or not. Future cars brought into America will be equipped with speedometers reading in miles.

The body was designed by Ghia, and it is a definite improvement over the current Sprite, although it must be remembered that the Sprite has come a long way in appearance since the original goggle-eyed little monster appeared. However, we felt that some of the Italian flair is lacking, and if the car was dressed up in one of those really exciting little coupe bodies, with perhaps a set of Borrani wheels, it would find a larger market in America at the price.

A welcome improvement over the BMC Sprite is the inclusion of windup windows, which are slightly curved to add elbow room to the interior. When the top is in position, the interior is free of drafts and the car can be used comfortably in all weather. Another good feature is a locking glove compartment to contain one's valuable possessions when the car is left unattended with the top down. Our test car was not equipped with a tonneau cover, but this is offered as an optional extra, and the only other items necessary to make the car completely civilized are safety belts and a radio.

Although the interior of the car is both attractive and well laid out, the seating accommodation does not measure up too well in our new "driver comfort rating," which will be included in the data panel of all our future road tests. For drivers up to about 5 ft 9 in. it is excellent, but the rating drops off sharply for taller drivers, so that six-footers and over will find it pretty snug. The reason for this apparently is that it is extremely difficult to design a sports car with such diminutive proportions which will allow the tall driver sufficient room for his feet, knees and elbows.

On the road, the performance is as spritely as could be expected. Our test car was "factory-fresh" when we picked it up, and the inherent stiffness of a new BMC product was apparent. However, we were able to drive the car for several hundred miles before carrying out our test, and both the engine and transmission seemed to free up noticeably during that period. With 58 bhp at 5700 rpm and 60 lb ft of torque at 3000 rpm, the power output is limited, but the weight of the car is only 1710 lb with driver so the performance is quite lively.

Taking the car over one of our test roads, which is situated in hilly country and rises to a summit of 2600 feet, we were surprised to find that we were able to reach the top without dropping down to third gear more than twice. This was achieved with a 2-man test crew and equipment, giving a total weight of about 1925 lb, and is a very creditable performance. However, the axle ratio is well suited to this type of driving because the engine is turning 3400 rpm at 50 mph, which is 400 rpm above the point of maximum torque. On the other hand, the situation is reversed under freeway driving conditions because the engine is turning at more than 4500 rpm at a cruising speed of 70 mph and this seems an excessively high engine speed for long periods of cruising.

The handling characteristics are almost identical to those of the Sprite, although the Pirelli Cinturatos on the Innocenti S seem to give it slightly better adhesion. The ride is comparatively hard and the combination of a short wheelbase and small wheels tends to make the car pitch on some surfaces. One can sail through fast turns at high speed without body lean and with the tail hung out slightly in true sports car fashion. The handling is practically neutral, with no

ROAD TEST

INNOCENTI S

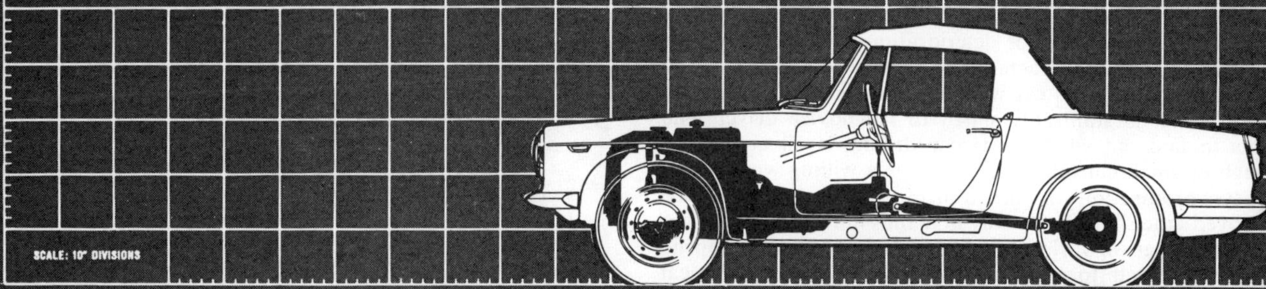

PRICE

List, West Coast POE	$2920
As tested, West Coast	$2937

ENGINE

Engine, no. cyl, type	4-cyl ohv
Bore x stroke, in.	2.54 x 3.30
Displacement, cc	1098
Equivalent cu in	66.98
Compression ratio	9:1
Bhp @ rpm	58 @ 5700
Equivalent mph	86.5
Torque @ rpm, lb-ft	60 @ 3000
Equivalent mph	46.5
Carburetor, no., make	2 SU
No. barrels—diameter	1-1.25 in.
Type fuel required	Premium

DRIVE TRAIN

Clutch diameter & type	6.25 in. single plate, dry
Gear ratios, 4th (1.00)	4.22
3rd (1.36)	5.73
2nd (1.92)	8.09
1st (3.20)	13.50
Synchromesh	2, 3 and 4
Differential ratio	4.22
Optional ratios	4.55, 3.73, 3.90

CHASSIS & SUSPENSION

Frame type	Platform
Brake type	disc/drum
Swept area, sq in	190
Tire size	145 x 13
Wheel revs/mi	941
Steering type	rack & pinion
Overall ratio	n.a.
Turns, lock to lock	2.3
Turning circle, ft	32

Front suspension: Independent with unequal length A-arms, coil springs and lever shocks.

Rear suspension: Live with leaf springs and tube shocks.

ACCOMMODATION

Normal capacity, persons	2
Hip room, in	2 x 18
Head room	36.5
Seat back adjustment, deg	0
Entrance height, in	42
Step-over height	14.5
Floor height	8
Door width	34
Driver comfort rating:	
for driver 69-in. tall	90
for driver 72-in. tall	55
for driver 75-in. tall	50

GENERAL

Curb weight, lb	1560
Test weight	1925
Weight distribution with driver, percent	51/49
Wheelbase, in	80
Track, front/rear	46.8/45.8
Overall length	136
Width	58
Height	47
Frontal area, sq ft	15.2
Ground clearance, in	5.5
Overhang, front	25.4
Rear	30
Departure angle, no load, deg	19
Usable trunk space, cu ft	3.1
Fuel tank capacity, gal	7.5

INSTRUMENTATION

Instruments: oil pressure, speedometer, tachometer, water temp, fuel gauge.

Warning lamps: high beam, lights, turn indicators, ignition, water temp, fuel level.

MISCELLANEOUS

Body styles available: roadster as tested.

ACCESSORIES

Included in list price: heater, windshield washer, cigar lighter, back-up lights.

Available at extra cost: tonneau cover, radio.

CALCULATED DATA

Lb/hp (test wt)	33.2
Cu ft/ton mi	73.7
Mph/1000 rpm (4th)	15.2
Engine revs/mi	3960
Piston travel, ft/mi	2155
Rpm @ 2500 ft/min	4580
Equivalent mph	69.2
R&T wear index	86.2

MAINTENANCE

Crankcase capacity, qt	4
Change interval, mi	3000
Oil filter type	full flow
Change interval, mi	3000
Lubrication grease points	12
Lube interval, mi	3000
Tire pressures, front/rear, psi	18/20

ROAD TEST RESULTS

ACCELERATION

0-30 mph, sec	5.6
0-40 mph	8.3
0-50 mph	12.6
0-60 mph	18.4
0-70 mph	28.0
0-80 mph	43.0
Passing test, 50-70 mph	13.9
Standing ¼ mi, sec	21.0
Speed at end, mph	63

TOP SPEEDS

High gear (5600), mph	85
3rd (6000)	67
2nd (6000)	47
1st (6000)	29

GRADE CLIMBING

(Tapley Data)

4th gear, max gradient, %	8.7
3rd	12.6
2nd	18.3
Total drag at 60 mph, lb	95

SPEEDOMETER ERROR

30 mph indicated	actual 28
40 mph	37.5
60 mph	57
80 mph	76

FUEL CONSUMPTION

Normal range, mpg	28–33
Cruising range, mi	210–245

ACCELERATION & COASTING

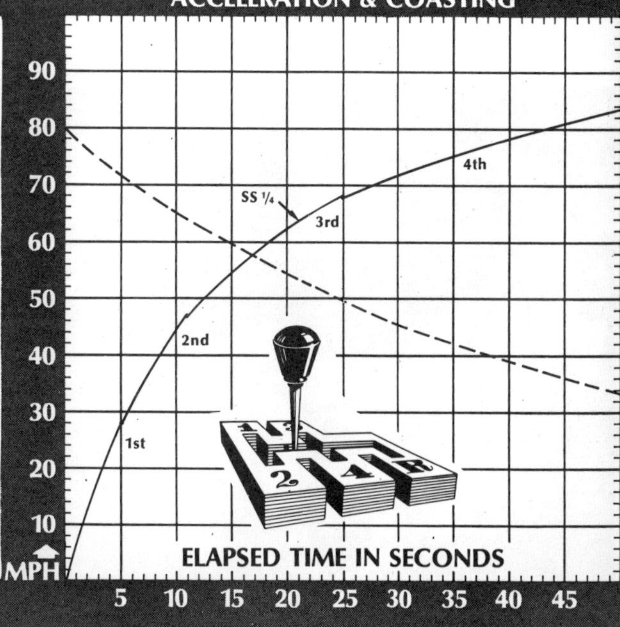

unfortunate tendencies at all, and the BMC rack and pinion with two and a quarter turns lock to lock insures positive and accurate steering.

To complement the road holding and steering, disc brakes are fitted at the front with drums at the rear, and this system seems to provide an almost indefinite number of high speed stops without complaint or deterioration in stopping ability.

These excellent characteristics make the Innocenti S a very good small sports car which accomplishes all it sets out to do, and it is admirably suited to the novice driver who wants to find out what sports car driving is all about. Although of small capacity, the engine is a willing worker and the synchronization of the transmission has been much improved since the Sprite was first introduced, but 1st gear is still of the crash variety. Over a period of years, the Sprite has proved itself to be a rugged, reliable and economical sports car to operate, and all the good features of the car are present in the Innocenti S.

However, assessing the relative merits of the Innocenti S presents a problem because it is necessary to take into account the price tag. At $2920 it is definitely out of the Sprite price range, and while it is certainly not inferior to the Sprite in any way, it is not $800 better and, furthermore, is not strikingly different. There is considerable pleasure to be derived from owning a car which is not frequently encountered, particularly if it bears the mark of an Italian coachbuilder. Unfortunately, the Innocenti S is not sufficiently individual to stand out very far in a crowd.

On the other hand, it is undoubtedly a well finished and attractively designed small sports car, and it has the added virtue of being built around the well-tried BMC components, which eliminates the parts and service problem associated with any limited-production automobile. For those people who are enamored with the Sprite but would like, and can afford, a little more comfort and a little more class, the Innocenti S may well be the answer.

[*"Driver comfort rating," a new addition to the R&T data panel, shows the percentage of driver comfort characteristics that met the standards described in "The 99th Man," R&T, January 1964. In general, the higher the rating, the more comfortable the car will be for the driver. Of all the sports cars measured so far, the average rating is 80 and 57, respectively, for drivers 69-in. and 75-in. tall. Thus a car with a rating of 90, 55 and 50 for 69, 72 and 75-in. tall drivers would be more comfortable than average for the driver of average height, and less comfortable than average for the tall driver.*]

AUSTIN-HEALEY 3000 MK III

The latest model has an additional 11 bhp, a bit of tarting up and still retains that classic cowl shake

TEN YEARS AGO, the Austin-Healey 100 was first introduced to the American market. The man behind it was Donald Healey, who had previously built a limited quantity of very advanced and successful sports cars using Riley engines. The demise of the Riley and the many problems confronting the small manufacturer caused Healey to seek an easier way of making a living, so he sold the Austin company on the idea of using production Austin parts to build a sports car aimed specifically at the American market. The result was the Austin-Healey 100, which has been gradually updated during the ensuing years to become the Austin-Healey 3000 Mk III Sports Convertible.

In its original form, the Healey was a conventional low-priced sports car, which offered the fast and responsive rag-top-and-flapping-side-curtains type of driving demanded by the enthusiast of the day. Its 4-cyl engine was lusty and reliable, its 3-speed and overdrive transmission was a novelty and, even if the car did have some extremely annoying faults, at least it provided enjoyable and exciting driving at a time when the Detroit product was nothing short of ghastly.

In its 1964 form, the Healey is a much more refined car although it retains many of the characteristics of the original version, both desirable and undesirable. The major changes consist of a general tarting up of the interior, which now has walnut veneer on the instrument panel and much more luxurious finish and detailing throughout. The top is good by sports car standards because it can be erected from inside the car, and it is secured by two over-center clamps on the windshield pillars. The windows are of the wind-up variety so that the car is well weatherproofed for the winter, at least in some climates.

BMC discarded the 4-cyl engine in 1956, substituting its rugged Six in the Healey, and this power unit has undergone a number of changes over the years so that it is now rated at 148 bhp. This is an increase of 11 bhp over the MK II version of the car, and the extra power is obtained from a different camshaft and the use of two 2-in. HD-8-SU carburetors. These carburetors require a lot of choke when starting from cold, and even then there is a tendency to spit back if too much throttle is used during the long warm-up period the engine seems to demand.

Another relatively insignificant but worthwhile improvement

AUSTIN-HEALEY 3000

AT A GLANCE...

Price as tested	$3828
Engine	6 cyl, ohv, 2912 cc, 148 bhp
Curb weight, lb	2650
Top speed, mph	116
Acceleration, 0–60 mph, sec	9.8
Passing test, 50–70 mph, sec	5.7
Average fuel consumption, mpg	19

AUSTIN-HEALEY 3000 MK III

concerns the rear suspension. Here the chassis frame has been curved to allow for additional axle movement, and springs of a lower rate are used. In order to control the movement of the axle more positively, it is now located by twin radius arms, and the results of these modifications are a better ride, improved handling and increased ground clearance.

Despite improvements to the rear suspension, the ride and handling of the Healey are still in the classic sports car tradition of the early '50s, before people like Chapman and Cooper had got in on the act. The ride is firm and there is some chassis flexing evident from the authentic cowl shake which occurs on any but the smoothest surfaces. In common with the rigidly sprung cars of the time, the road-holding of the Healey is quite good on smooth surfaces, but a fast turn on a poor surface may produce surprises, because the rear end has a tendency to jump and skip over bumps.

Under normal conditions, the Healey has slight understeer, but plenty of throttle opening in a corner will change this to oversteer, and the tail can be hung out a long way with delicate throttle control. However, one should select the road surface rather carefully before experimenting because any irregularity will throw the car off its line, and a wet surface tends to make the handling treacherous.

At low speeds the steering is comparatively heavy, but it lightens as the speed increases and the car holds its line accurately at normal cruising speeds, although there is a tendency for the rear end to twitch on rougher surfaces.

Due to the improved rear suspension and the slight rearward weight bias, one can apply a surprising amount of power when coming out of a slow turn, without any tendency for the car to break away, lift its inside rear wheel, or lose traction. The Austin-Healey is no sluggard on the road, but one has to work hard and use some skill and experience to extract the maximum from it.

Although the car has an adjustable steering column, it is difficult for tall drivers to accommodate themselves comfortably, as a glance at the Driver Comfort Rating on the data panel will show. Apart from the driver and passenger seats, there is accommodation for two children behind, or, alternatively, one adult can be carried for short trips. This space can also be used to supplement the limited trunk space, which is almost entirely filled by the spare wheel and the battery. The controls are a mixture of ancient and modern. The transmission, in particular, needs updating because 1st is unsynchron-

ROAD TEST

AUSTIN-HEALEY 3000

PRICE

List price	$3699
Price as tested	$3828

ENGINE

No. cylinders & type	6 cyl, ohv
Bore x stroke, in.	3.28 x 3.50
Displacement cc	2912
Equivalent cu in	177.7
Compression ratio	9:1
Bph @ rpm	148 @ 5250
Equivalent mph	123
Torque @ rpm, lb-ft	165 @ 3500
Equivalent mph	82
Carburetors, no. & make	2 SU
No. barrels & dia	1–2 in
Type fuel required	premium

DRIVE TRAIN

Clutch type	single plate, diaphragm
Diameter, in	9.5
Gear ratios: o'drive (0.82)	3.21:1
4th (1.00)	3.91:1
3rd (1.31)	5.12:1
2nd (2.06)	8.05:1
1st (2.88)	11.3:1
Synchromesh	on top 3
Differential type	hypoid bevel
Ratio	3.91:1
Optional ratio	3.55:1

CHASSIS & SUSPENSION

Frame type	separate ladder type
Brake type	disc/drum
Swept area, sq in	383
Tire size	Dunlop RS-5 5.90 x 15
Steering type	cam & lever
Overall ratio	15:1
Turns, lock to lock	3.0
Turning circle, ft	36.0

Front suspension: independent with A-arms, lever shocks, coil springs.

Rear suspension: live axle with semi-elliptic springs, lever shocks, Panhard rod.

ACCOMMODATION

Normal capacity, persons	2
Occasional capacity	3
Seat width, front, in	2 x 17.5
Rear	2 x 16
Head room, front/rear	35.5/28.0
Seat back adjustment, deg	none
Entrance height, in	44
Step-over height	14.5
Door width, front	26.5
Driver comfort rating:	
For driver 69-in. tall	85
For driver 72-in.	65
For driver 75-in. tall	45

(85/100, good; 70/85, fair; under 70, poor)

GENERAL

Curb weight, lb	2650
Test weight	3020
Weight distribution (with driver), front/rear, %	47/53
Wheelbase, in	92.0
Track, front/rear	48.7/50.0
Overall length, in.	157.0
Width	60.0
Height	50.0
Frontal area, sq ft	16.7
Ground clearance, in	4.5
Overhang, front/rear	25/41
Departure angle (no load), deg	14
Usable trunk space, cu ft	2.3
Fuel tank capacity, gal	14.4

INSTRUMENTATION

Instruments: fuel, oil pressure, water temperature, 140 mph speedometer, 7000 rpm tachometer.

Warnings lights: high beam, ignition, turn signals.

MISCELLANEOUS

Body styles available: roadster as tested.

ACCESSORIES

Included in list price: overdrive, wire wheels, heater, windshield washers, full instrumentation, adjustable steering column, seat belt anchors.

Available at extra cost: seat belts, tonneau cover.

CALCULATED DATA

Lb/hp (test weight)	20.4
Cu ft/ton mi	86.7
Mph/1000 rpm (overdrive)	23.3
Engine revs/mi	2560
Piston travel, ft/mi	1490
Rpm @ 2500 ft/min	4280
Equivalent mph	100.3
R&T wear index	37.9

MAINTENANCE

Crankcase capacity, qt	7.5
Change interval, mi	3000
Oil filter type	full flow
Change interval, mi	6000
Chassis lube interval, mi	3000
Tire pressure, front/rear, psi	20/25

ROAD TEST RESULTS

ACCELERATION

0–30 mph, sec	3.6
0–40 mph	5.1
0–50 mph	7.0
0–60 mph	9.8
0–70 mph	13.0
0–80 mph	17.1
0–100 mph	27.4
Passing test, 50–70 mph	5.7
Standing ¼ mi	17.4
Speed at end, mph	82

TOP SPEEDS

Overdrive (5000), mph	116
4th (5500)	101
3rd (5500)	81
2nd (5500)	51
1st (5500)	36

GRADE CLIMBING

(Tapley data)

Overdrive, max gradient, %	9
4th	12
3rd	17
2nd	27
1st	off scale
Total drag at 60 mph, lb	95

SPEEDOMETER ERROR

30 mph indicated	actual 30.0
40 mph	40.2
60 mph	60.0
80 mph	79.1
100 mph	98.7

FUEL CONSUMPTION

Normal driving, mpg	16/21
Crusing range, mi	230/302

ACCELERATION & COASTING

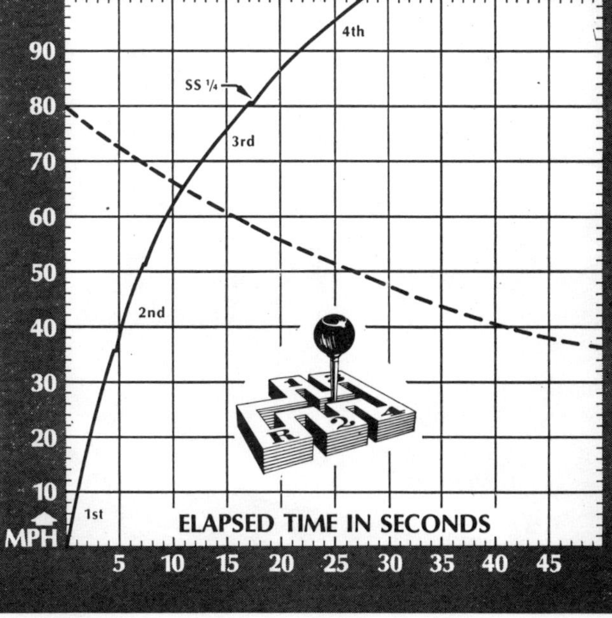

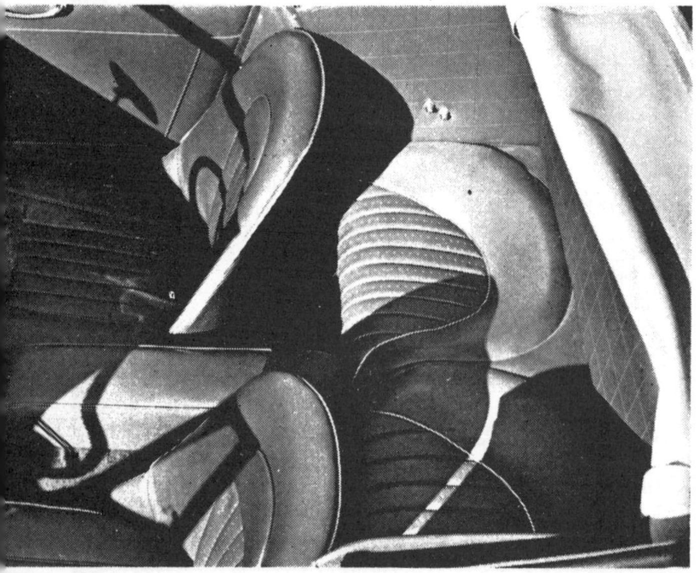

ized, noisy and difficult to select, and the synchronization of the other three gears leaves a lot to be desired. On the other hand, the shifting of the overdrive is very fast and smooth, although the positioning of the control switch makes it hard to locate in a hurry.

The overdrive operates on 3rd and high so that in effect the car has 6 speeds. Although this would seem to provide a ratio for any occasion, there is actually an excessive gap between 2nd and 3rd, which is accentuated by the fact that if the overdrive switch is in the overdrive position, the overdrive ratio will be selected automatically as soon as one shifts up from 2nd to 3rd, thereby increasing the gap.

The clutch is a great improvement over previous Healey models and, in fact, over the majority of sports car clutches. It is of the diaphragm type, and its action is light and smooth in direct contrast to the early Healey units, which were very heavy and fierce, and had an annoying habit of breaking their linkages. The car is equipped with disc brakes at the front and drums at the rear and the brakes are generally very good, although the pedal pressure required is high despite the use of a booster in the system.

One of the most noticeable features of the Healey is the unduly high noise level, which commences at about 3000 rpm and makes normal conversation impossible by 4000 rpm. It takes the form of a loud roar which is a mixture of engine, fan, transmission, carburetor, and exhaust noises, and one can't help feeling that a lot of it could be eliminated by better attention to insulation. The noise coupled with the rather uncomfortable driving position and the stiff suspension makes the car tiring to drive on a long trip. The level of exhaust noise has been reduced by the addition of a pair of supplementary mufflers located transversely at the rear of the car. However, the eternal Healey problem of low ground clearance remains, and the primary mufflers are still located beside the left frame rail where they are far too low and consequently extremely vulnerable.

The Austin-Healey is now in its 12th year, which is a long span in an age of rapid automobile development. It has served its purpose well both on road and track, and as recently as 1962 a Healey got as high as 8th place at Le Mans before piston failure put it out. However, the existing model appears to have reached the point of honorable retirement, and perhaps BMC will soon come up with a worthy successor after starting from scratch with a clean sheet of paper.

Sebring Sprite Prototype

BY DAVID PALMER

TWO INTERNATIONAL EVENTS on the racing calendar that the British Motor Corporation Department always tries to support are the classic sports car races at Le Mans and Sebring. While they have met with a modicum of success at the French race, Sebring has generally seemed to turn out as little more than an excuse for showing the BMC flag in the U.S. This year, however, a 1.3-liter Austin-Healey Sprite prototype driven by Clive Baker and Rauno Aaltonen scored a resounding success. This sleek little coupe, going like lightning in the rain, compared with almost everything else on the wet circuit, won its class, finished ahead of the fastest of the Alfa Romeo TZ-Cs, beat its 3-liter big brother A-H 3000 and occupied a proud 15th overall in the final standings.

There were two of the 1.3-liter Sprite prototypes in the race this year. The second car, driven by Paddy Hopkirk and Timo Makinen, finished 18th at Sebring, likewise a thoroughly respectable position in such quick company. The two cars were designed jointly by Donald Healey Motors at Warwick and BMC. Although they differ slightly in bodywork, both are roughly the same under the skin.

Although in appearance the prototypes bear no resemblance at all to the current Mk III Sprite, they are surprisingly basic underneath, the extra power and performance being obtained partly from the larger engine and less weight and partly from very careful preparation.

The body was developed from an original Healey idea, made as a mock-up and tested in the BMC wind tunnel before the actual final version was made by hand from aluminum and mounted on a Sprite floor pressing. The nose tapers down to a sharp edge, with the twin lamps and flashing signals enclosed in perspex molded to the body line. All windows are of perspex to save as much weight as possible, those in the doors sliding down for ventilation. There is no interior trim; the fascia is finished in matt black, with the main instruments situated in a nacelle directly ahead of the 3-spoke, leather-grip steering wheel. To the left of the nacelle are flick switches for lights, wipers, washers, and an overriding switch for the fuel pump. The speedometer is ahead of the passenger seat and a fire extinguisher is mounted in a clip on the gearbox tunnel.

The top of the car sweeps gently down from the windscreen with the tail suddenly chopped off flat in the current idiom. The space thus achieved is filled mainly with a 20-gal. fuel tank that gives rise to a high floor line inside. The spare wheel is the only adornment to this floor. Two racing bucket seats complete the interior, which is decidedly cramped, and forced BMC team chief Stuart Turner to allocate the smallest of his drivers, Clive Baker and Rauno Aaltonen, to this car.

Motive power for the car is a rather special version of the BMC "A" series engine; its nearest similar unit in production is that used on the Mini Cooper S. The 1275-cc block is overbored 0.020 in. to give a new capacity of 1293 cc, and is a conventional engine with sump, not a front-wheel-drive package. The crankshaft is balanced and runs in normal Cooper S bearing shells and the rod bearings are also production, but the camshaft is a competition version having higher lift (0.394-in.) and a 95° overlap. The inlet opens 50° before top dead center and closes 70° after bottom dead center. Exhaust opens 75° bbdc. and closes 45° atdc. The rockers are lightened and the pushrods matched for weight.

A Cooper S cylinder head is used, with standard-sized valves, but the combustion chambers are gas-flowed and the porting matched perfectly to the manifolds. Machining of the head and block has raised the compression ratio to something in the region of 12:1 and as an extra safety precaution forged pistons instead of cast ones are used, although they are made from the same basic material as the production Cooper S pistons. With a static ignition timing of 5° btdc and Champion N58R plugs, maximum rpm is between 7000 and 7250.

Special manifolding, allowing the most direct route to and from the combustion chambers without resorting to the crossflow method, carries a twin-choke Weber 45 DCOE to get the fuel in and a 3-branch free-flow exhaust, connected

Radiator air intake is below British number plate.

Cockpit has only necessary equipment.

Tail is chopped off in current fashion prescribed by Dr. Kamm.

to a straight-through pipe with barely any silencing qualities, takes the burned gas away. At maximum revs the engine develops 110 bhp, while maximum torque of 94 lb-ft comes at 5000 rpm. Oil pressure at maximum revolutions is between 80 and 85 psi at a temperature of 80° C.

Transmission is as the standard Sprite except for changes to ratios. A Borg & Beck diaphragm spring clutch with competition linings on the driven plate transmits power to a 4-speed close-ratio gearbox, the solid rear axle having a 3.9:1 final drive unit.

Lockheed disc brakes of 7-in. diameter are fitted to the car all round and the disc pads use Mintex lining material, but there is no power assistance. The hubs have four studs to carry special Healey built 13-in. cast magnesium wheels with eight large cooling ducts in each. The tires are 5.00L-13 Dunlop Racing. Use of these special wheels makes a saving of something like 4 lb in unsprung weight per wheel. Apart from alterations to the shock absorber valves and the addition of an anti-roll bar across the front, the suspensions remain exactly as per the production car.

It can be seen from this description that the car does not differ as much as one would think by its appearance from the current Mk III Sprite. Probably the only reason it had to be entered in the prototype class was so that the large engine could be used, and once this had been decided the special light-weight body and other advantageous modifications could be made. Unfortunately, just because a 1.3-liter Sprite is being raced as a prototype is no guarantee that a production model will follow in due course, although there is always a possibility of this happening.

Sebring co-driver Clive Baker about to set off on a trial run.

LEO BESTGEN ILLUSTRATION

Once upon a time, far away on the misty distant island where King Arthur once held sway, a humdrum little sedan was chugging its way into blissful anonymity. Then, one day, a wizard named Donald Healey came along and waved a magic wand at it, and created—presto—

SPRITE: the enchanted frog

BY JACK P. MALONEY

ACTUALLY, THE SPRITE just *looked* like a frog. Deep down inside, it was one of the most distinctive, commonplace, homely, attractive, antiquated, modern, irritating and enjoyable sports cars ever to hop merrily onto the scene. From the day it first appeared on the market in 1958 until it was replaced by the Sprite Mark II in 1961, the "frog-eye" Sprite had a greater impact on the growth of motor sports in the United States than any car since the MG TC.

Donald Healey's design caught the interest of the British Motor Corporation in 1957. The advantages of the Sprite idea were many: most of the mechanical bits and pieces came from BMC cars already in production: it was cheap and simple to build; it would reach a low-price market where there was no competition; and the design looked good.

At least, the original design looked good, for Healey's brainchild included retractable headlights, similar to those in today's Lotus Elan. However, the budgeteers at BMC decided that the cost of retractable headlights was too high, so they simply left them up in the middle of the hood. Thus the most distinctive—and homely—aspect of the Sprite's appearance was born.

The rest of the body incorporated unusual features, too. The whole front end—fenders, grille and all—opened up to allow easy access to the engine and front suspension. The trunk space was reached through an opening behind the seats. This gave a tremendous amount of room for odd-shaped luggage if the owner was agile enough to get items in and out, and also contributed greatly to the fortunes of chiropractors and spine specialists.

Despite the Sprite's small size, it had more shoulder and leg room than the MGA, due to its almost completely hollow construction. No attempt was made to hide the box-section frame members, no heavy padding covered panels, no stylish fender lines bulged out from the doors. It was a no-nonsense car, closer to a lightly-bodied go-kart than a sophisticated automobile. If it drummed and echoed a little, or seemed excessively Spartan, well, at least nothing was wasted in the design. And the price—under $2000 complete—was right.

The running gear on the Sprite showed more economy than imagination. The engine was the long-stroke 948-cc "A" engine used in several of BMC's smaller sedans, the only changes being in the valves, valve springs, bearings, and the use of two 1⅛-in. SU carbs. The gearbox, with its lamentable gap between second and third gears, was unchanged. Steering and front suspension came from other BMC sedans, and the rear suspension featured rather hoppy quarter-elliptic springs. The brakes, 7-in. drums straight from the Morris Minor 1000, were just adequate for normal driving, and marginal at best for serious competition.

If the drive train of the Sprite was uninspired, the straight-line performance rather modest and the styling somewhat bizzare, what accounts for the fact that about 40,000 were sold in this country during its three years of manufacture, and that thousands are still regularly used for competition? Two features endeared the Sprite to the public: it was cheap, and it was fun to drive.

The tight, light steering of the Sprite, its almost neutral cornering and forgiving nature made it a perfect car in which to learn about driving. Many's the driver who recovered from tight situations in a Sprite who would have been in serious trouble in any other car.

To the average sports car driver who began in, say, an MGA or Triumph and never looked back, the Sprite was just a cheaper, smaller and slower car. Vast numbers of enthusiasts who have driven almost all of the mass-produced sports cars have never bothered to try a Sprite . . . much to their loss. For sheer effortless, seat-of-the-pants driving, immediate reponse and absolutely predictable cornering, it was unmatched among mass-produced cars of its day.

That's not to say there weren't drawbacks to the frog-eye Sprite. The earliest models had snap-fasteners for attaching the top to the windshield, and driving above 40 mph on a rainy day resulted in getting more water on the inside of the windshield than the outside. The 1958 Sprites also had the coil mounted on the generator, and engine vibration almost invariably shook the coil loose and ruined it. These two features were corrected within the year, but the car was hardly perfected.

The top, as the owner's manual so brightly put it,. was "erected." This meant that two people took several minutes to assemble, snap, stretch, clamp and screw the top and side-curtains on . . . usually in the middle of a rainstorm. As a

SPRITE:

result, Sprite owners tended to drive *al fresco* under the worst weather conditions rather than struggle with the fiendish top. One needed good health to own a Sprite.

The 948-cc engine tended to leak oil from every conceivable, and many an inconceivable, place. And if the tiny by-pass water hose between cylinder head and block ever started leaking, the owner could count on three hours of struggle, several banged knuckles, and a thorough review of his scatological vocabulary before he could get a new one installed. The ignition system was highly vulnerable to moisture and wiser owners relied on silicone grease to protect the wiring.

Another chronic problem with the Sprite was the mechanical tachometer drive. In cold climates, the small (and expensive) gear reduction box between the generator and tach cable could be counted on to break every winter. Sprite owners had the choices of (a) disconnecting the tach every winter, (b) buying an electric tach, or (c) joining thousands of other Sprite owners every spring in waiting for back-ordered tach drive boxes.

If the Sprite called for frequent mechanical attention, it was at least well designed for maintenance. With the massive hood propped up, almost everything of importance on the engine and gearbox was easily reached. And if major surgery was required, two healthy men with a chain and a long iron pipe could hoist out engine and gearbox together in short order. The absence of fixed bodywork in front of the engine made lifting almost unnecessary.

The advent of the Sprite put new life into SCCA H Production racing, and even today HP races bring a gaggle of goggle-eyed Sprites burbling around tracks acoss the country. Simple maintenance and the ready availability of cheap parts make the Sprite one of the easiest cars to campaign.

In the international field, the Sprite has had a good share of success despite the fact that the British Motor Corporation has done most of its competition work with the big Healeys and, lately, the hot Minis. In less than three years, Sprites recorded ten class wins in the Alpine, Circuit of Ireland, Scottish, Liege-Rome-Liege, R.A.C., Portuguese and Tulip rallies, usually finishing well up in the overall results. Probably the most successful Sprite competitor was England's John Sprinzel, who took four rally wins and numerous international racing honors with his specially built Sprinzel Sprites.

The Sprite's racing abilities were proven by class wins in international competition at Sebring, Le Mans, Montlhéry, Goodwood, Aintree, Brands Hatch, Nassau and Nürburgring. Even drivers such as Stirling Moss, the late Walt Hansgen, Briggs Cunningham and Bruce McLaren took turns at the wheel; Moss drove Sprites in two Sebring 4-hour races, taking his class in 1960 and finishing behind Hansgen and McLaren in 1961. A special-bodied, supercharged Sprite set 50 National and 15 International Class G speed records at Bonneville in 1959, and a Speedwell Sprite driven by Graham Hill and George Hulbert set new Belgian National Class G records in 1960.

This kind of class-winning, record-setting performance isn't remarkable in the ever changing world of motor sports; many other sports cars can boast of even more impessive records. But the Sprite's record is significant for two reasons: first, because it was designed, built and marketed as a "poor man's" sports car; second, because most of its successes came, not under the auspices of a full-blown factory competition program, but in the hands of private owners.

In May 1961, after almost 49,000 odd-looking, homely frog-eyed Sprites had rolled off the BMC assembly lines, the Sprite Mark II was introduced. Mechanically similar to its predecessor, the Mark II actually was a completely different car. Gone was the massive one-piece front end that made maintenance so easy; gone the peculiar "tucked-in" tail and awkward luggage access; and gone the freakish headlights and smiling grille that distinguished Sprites from every other car on the road. In their place was a commonplace "modern" car with a normal trunk lid, fixed front fenders, ordinary headlights, unimaginative grille and conventional hood. No one could call the Mark II freakish or ugly, or laugh at the design, but no one could say much of anything else about it, either.

But the Sprite—the car produced from 1958 to 1961 and never designated "Mark I" while it was in production—was never a car to compare to anything else. If the Sprite was homely, it was never commonplace. If it was Spartan, it was also cheap. And if it was modestly powered, it at least set a standard for *responsive* handling that was unmatched in its time. The Sprite helped introduce tens of thousands of Americans to motor sports, because its price was within reach of people who might never have had a sports car otherwise. It created a market where none had been before; for this reason alone, the Sprite, frog eyes and all, deserves a niche in modern sports car history side by side with the MG TC.

GORDON CHITTENDEN PHOTOS

ROAD & TRACK COMPARISON TEST
4 BASIC SPORTS CARS

AUSTIN-HEALEY SPRITE DATSUN 1600 FIAT 850 SPIDER TRIUMPH SPITFIRE

ALPHABETICALLY, the four cars we chose for comparison in this article are the Austin-Healey Sprite (which is the same as the MG Midget except for trim and the badge), Datsun 1600 Sports, Fiat 850 Spider and Triumph Spitfire. Although they are not all exactly in the same group in all characteristics (the engines range from 817 cc to 1595; the prices from $2100 to more than $2700), they are the four major 2-seater sports cars in the lowest-price group and as such deserve to be looked at together.

Because these basic sports cars are most likely to be strongest in appeal to the younger driver, one who has never owned a sports car (or perhaps any car) before, we called in our resident under-voting-age enthusiasts, Jay and Mike, to assist us in our evaluation. They are both college students; one drives a stock 1959 VW Beetle, the other a highly modified 1962 Sprite with rollbar and rally stripe, and both are much enamored of automobiles. They drove all four of the cars, put them through their paces on a tight little handling course we set up on a parking lot at Orange County International Raceway and also put in a few highway and freeway miles as well. In the following evaluation, we have relied heavily on their opinions to tell us what they—and by extension what prospective owners in their age group—find that adds or detracts from each car's appeal.

In general, they see the basic sports car as a "beginner" car for many 16-to-20-year-old drivers; the first car many of them will own, the car they will be looking forward to as soon as they get their driver's licenses. One of the most appealing features of these cars is cost, as might be expected. These are the lowest priced sports cars on the market; cheapest in original outlay, moderate in operating cost, relatively inexpensive to keep in repair. Also of fundamental importance is the fact that they are far more likely to receive parental approval than either an in-car like a wide-tracked, drag-slicked Musclecar or a more powerful sports car. "My folks didn't bat an eye when I talked about getting a Fiat," Jay says, "But when I mentioned Porsche, boy, they went right to the ceiling." It is also of some importance that insurance companies are slightly less appalled at young drivers in basic sports cars than in machines with super power.

The fact that small sports cars of this type are socially acceptable is also basic in the appeal. The competition image is

Fiat 850 Spider, with lowest price and least power, was judged to be prettiest of the four basic sports cars. A bolt-on hardtop (below) was recently introduced in the U.S.

Datsun 1600 has most performance and largest price tag.

Spitfire's rear end adopts this attitude in hard cornering.

PERFORMANCE

	Sprite	Datsun	Fiat	Spitfire
Time to distance, sec				
0–100 ft	4.0	3.8	4.7	3.8
0–500 ft	10.5	10.8	11.7	10.3
Standing ¼-mi	19.9	19.9	21.7	19.3
Speed at end, mph	69	70	62	70
Time to speed, sec				
0–30 mph	4.3	3.8	5.7	4.1
0–45 mph	8.0	7.9	11.1	7.4
0–60 mph	14.7	13.3	20.0	13.6
0–75 mph	25.6	25.0	38.0	22.6
Top speed, mph	93	101	84	100
Overall brake rating	good	very good	good	very good
Speedometer error:				
When speedo reads 60, actual speed is	58.0	58.4	57.5	57.5

4 BASIC SPORTS CARS

important in this respect and, as a consequence, the Sprite and Spitfire both rank high because of their long-standing success in amateur sports car racing. The Sprite also enjoys additional prestige as it is a popular slalom or autocross competitor—and these are competitive events in which the young driver can participate. The Datsun also enjoys some of the glamor rub-off from racing but judging from our resident experts, there is far less age-group identification with the Datsun. "Sprite drivers are a lot younger than Datsun drivers," they seem to say, and regard the Datsun as half a notch removed from the others because of its greater initial cost and bigger size. The Fiat 850 Spider is viewed as a bit too new yet to have established as strong an acceptability quotient as the Sprite and Spitfire.

Another of the important criteria established for these basic sports cars is that they are capable of being driven easily and well. This is perhaps an unexpectedly sophisticated requirement for a new driver but it is an extremely logical one as it implies an appreciation of predictable handling, accurate steering, good brakes, responsive controls and so on. Obviously, no driver is better than the potential of his car but the better the car, the more exciting and satisfying it can be to drive it well. By this criterion, all our test cars scored very high as they are all extremely easy to get used to and to feel comfortable driving.

Also of importance in the appeal of the basic sports car is the availability of information and parts for improving the performance. "Everybody wants his car to be a little faster and handle a little better than somebody's else's," Mike says. In this respect, the Sprite, Datsun and Spitfire owners are better off as all three have a very wide range of not-too-expensive performance-improving equipment available; of equal importance, there is good super-tuning information available. The Fiat suffers a little in comparison: its engine is already in a fairly high state of tune (52 bhp from 817 cc) and there isn't much equipment for it on the U.S. market—at least not yet.

Our young experts also pointed out a couple of areas that they would consider less important than would older drivers. Luggage space is not considered all that important, they told us; the young driver isn't likely to worry about it if there isn't room for a set of hard-sided matched suitcases. We also learned that the basic sports car is considered to be a roadster rather than a convertible. That is, there should be a top that can be put up for occasional weather protection but as the top is far more likely to be down than up, the young driver is more

SPECIFICATIONS

	Sprite	Datsun	Fiat	Spitfire
Engine type	ohv 4 cyl	ohv 4 cyl	ohv 4 cyl	ohv 4 cyl
Bore x stroke, mm	70.6 x 81.3	87.2 x 66.8	63.9 x 63.5	73.7 x 76.0
Displacement, cc	1275	1595	817	1296
Bhp @ rpm	65 @ 6000	96 @ 6000	52 @ 6200	75 @ 6000
Torque @ rpm, lb-ft	72 @ 3000	103@4000	45.6@4000	75 @ 4000
Gearbox type	4-spd man	4-spd man	4-spd man	4-spd man
Synchromesh	on top 3	on all 4	on all 4	on top 3
Overdrive available	No	No	No	Yes
Brakes, front/rear	disc/drum	disc/drum	disc/drum	disc/drum
Swept area, sq in	190	310	205	205
Tire size	5.20-13	5.60-14	5.50-13	5.20-13
Steering lock-to-lock	2.3	2.5	3.5	3.8
Curb weight, lb	1560	2085	1640	1680
Weight distribution, f/r	50/50	51/49	40/60	54/46
Wheelbase	80.0	89.8	79.8	83.0
Track, front/rear	46.3/44.8	50.0/47.1	47.1/49.2	49.0/48.0
Length	137.4	155.6	148.9	147.0
Height	48.6	51.4	48.0	47.5
Width	56.6	58.9	59.0	57.0
Frontal area, sq ft	15.3	16.8	15.7	15.1

Sprite, with moderate roll oversteer, has excellent handling.

CALCULATED DATA

	Sprite	Datsun	Fiat	Spitfire
Lb/bhp (test wt)	29.4	25.5	37.7	27.0
Mph/1000 rpm	15.4	18.4	13.6	15.9
Revs/Mi (60 mph)	3900	3260	4425	3760
Wear Index	81	47	82	71
Brake swept area/ton	200	253	209	202

likely to judge the car on how it looks without the top than how simple it is to erect.

Austin-Healey Sprite

IN YEARS OF service, the Sprite is the senior sports car of those tested and now is somewhat dated in its design. The engine, gearbox and running gear are all simple and straightforward. In size, it is the smallest and lightest of the four cars in this evaluation. The construction is solid and rattle-free and there's nothing cheap or flimsy about it. The weight distribution is exactly 50/50 and all our drivers agreed that its handling was excellent. There is a moderate quantity of roll oversteer built into the live rear axle suspension which tends to bring the rear end out when cornering vigorously but this is easily controlled and does not detract from the virtues of the Sprite's handling. The steering is excellent and combines just the right amount of quickness and sensitivity. Though both the Datsun and Spitfire are faster than the Sprite, all our drivers ranked the Sprite as superior in handling.

The Sprite also has several limitations resulting from its compact dimensions. First, the small size of the cockpit area almost automatically eliminates the 6-ft-or-over driver. The seat won't go back far enough to give sufficient leg room, the pedals are too small and too close together for any but the daintiest feet, the steering wheel is too close to the chest and the doors are small enough to make it difficult to get in or out, as well as having such a high windowsill as to aggravate claustrophobia.

And there are other faults that make it less pleasant than it could be. The manufacturer's solutions to the problems created by the U.S. safety regulations are poorly worked out, for example, and result in an ugly "safety car" look to the dashboard, an awkwardly hard-to-operate location for the ignition key and door latches that are almost impossible. It seems that the manufacturer must have said, "Well, if the Americans insist on these things, we'll make them just as ugly and inconvenient as possible." The Fiat and Datsun people solved the same problem with far more grace.

Other notes we made about the Sprite include the fact that the seats are far more comfortable than Sprite seats used to be, that the ride is softer and the handling better. We also noted that there was no locking compartment in the cockpit where valuables (or tape decks) might be locked away out of sight, that there were no armrests and that first gear was not synchronized in the otherwise excellent gearbox.

Summing up, we would say that because of the Sprite's excellent handling and straightforward mechanical specifica-

INSTRUMENTATION	Sprite	Datsun	Fiat	Spitfire
Speedometer, mph	100	120	120	120
Tachometer, rpm	7000	7000	8000	6500
Odometer, mi	99,999	99,999	99,999	99,999
Trip odometer, mi	999.9	999.9	999.9	999.9
Oil pressure gauge	Yes	Yes	Yes	No
Water temperature gauge	Yes	Yes	Yes	Yes
Ammeter	No	Yes	No	No

Like everything else about the Sprite, engine is tight fit.

Datsun's 96-bhp engine is largest, strongest of those tested.

Fiat engine, 52 bhp from 817 cc, is in fairly high state of tune.

Entire nose of Spitfire swings up to offer ultimate in access.

4 BASIC SPORTS CARS

tions, it would be hard to find a car that we could recommend more highly for the young driver who is interested in learning the good things there are to know about sports cars.

Datsun 1600 Sports

THE DATSUN 1600 Sports is the largest, heaviest, most powerful and most expensive of the four cars considered. It is similar in basic layout to the traditional British sports car. The handling is good but compared to the Sprite, which has similar suspension, the Datsun has more initial understeer. This characteristic, which results in the front end tending to stay on its original course when the car is cranked into a turn, makes it more difficult to tweak out the rear end so that it may be guided by a combination of steering and throttle like the Sprite. In any case, it's an easy car in which to get comfortable, both in seating position and in driving. It does not corner as flat and positively as the Sprite or Fiat and on tighter turns there's enough torque to lift an inside rear wheel, which neither the Sprite nor Fiat tend to do. This lift-and-spin characteristic also limits the rapidity with which the Datsun can be gotten underway and results in acceleration times which are not as quick as might be expected. In top speed, it is the fastest of the four.

In appearance, the Datsun is undistinguished, neither outstandingly ugly nor fetchingly pretty. One of our drivers decried the use of whitewall tires and several pointed out that tires with more traction would make a big improvement in handling.

The Datsun's list price includes several items that cost extra on the other cars. There is a nice radio, for instance; cigar lighter, whitewalls, tonneau cover and so on. The instrumentation is the most complete of the four cars and also includes a clock at no extra charge. There is also a locking map box between the seats, a locking glovebox on the dash, a courtesy light, scuffplates on the insides of the doors, synchromesh first gear and more horsepower and torque.

As noted earlier, the Datsun is generally considered to be an "older person's" car—and by "older," our young experts mean older than the typical owner to whom the Sprite, Fiat or Spitfire would appeal. Perhaps this partially results from the fact that it does not have the lively responsiveness of the

ACCOMMODATION & CONVENIENCE

	Sprite	Datsun	Fiat	Spitfire
Driver Comfort Rating (Scale of 100)				
Driver 69 in. tall	70	75	85	85
Driver 72 in. tall	60	70	75	75
Driver 75 in. tall	45	65	70	70
Seatback adjustment, degrees	0	10	0	0
Visor Mirror	No	No	Yes	Yes
Locking glovebox	No	Yes	Yes	No
Vent wings	Yes	No	No	No

PRICE & EQUIPMENT

	Sprite	Datsun	Fiat	Spitfire
List price	$2130	$2766	$2126	$2285
Price with similar equipment (radio, tonneau, heater, belts)	$2220	$2766	$2176	$2406
Extras Available:				
Hardtop	Yes	Yes	Yes	Yes
Overdrive	No	No	No	Yes
Mag wheels	Yes	Yes	Yes	Yes
Performance equip.	Yes	Yes	No	Yes
Workshop manual	Yes	Yes	Yes	Yes
Competition tuning manual	Yes	Yes[1]	No	Yes[2]
Basic maintenance interval, mi.	3000	3000	3000	3000
Warranty, mo/mi	12/12M	12/12M	12/12M	12/12M

[1]Datsun tuning manual available from Bob Sharp Racing, Wilton, Conn. $5.

[2]Spitfire tuning manual available from TR Competition Dept., 1957 W. 144th St., Gardena, Calif. 90249. $2.

Sprite's cockpit is smallest.

Datsun's most luxurious.

Fiat's best looking.

Spitfire's most comfortable.

smaller, lighter cars as well as the fact that its more luxurious appointments and fuller range of equipment make it seem a long way removed from what might be considered a lean and hungry sports car.

As we've said before, dollar for dollar, the Datsun 1600 offers more for the money than any other sports car. Yet for an audience for whom it should seem ideal, it's a bit fancier and a bit too expensive to hit the center of the market target. It seems that a starker, more nearly bare-bones model would increase its appeal with the younger market.

Fiat 850 Spider

ALTHOUGH THE Fiat 850 Spider has the smallest engine of the four cars in this comparison and offers the least in the way of performance, it is also the prettiest of the lot and has the smallest price tag. In comparison, the others look like older designs that have been revised to keep them somewhere near up to date while the Fiat is sparkly new in every respect. The body shape is outstanding, a range of bright, modern colors is offered and, accepted for what it is, the 850 is impossible to fault.

The suspension, independent at both ends, is well suited to the job it is asked to do. In cornering there's no feeling of rear-end heaviness even though 60% of the weight is on the rear wheels; there's no jacking or wheel-lifting in tight turns and when one presses very hard to force it into an oversteer attitude, the transition is smooth and easily controlled. The tail will swing out if you cut off power in the middle of a hard turn but even then, if you are aware of what's happening, it is easy to catch.

The body styling is by Bertone and says volumes for having your car styled by the best stylist you can hire. The design is superbly worked out, even to the revisions of the front fenders required by the safety regulations which now forbid the use of a cover over the headlights and to the metal cover under which the top is stowed. All the other cars have fabric boot covers, some already had missing snaps; the ease with which the Fiat top went up and down made the others seem clumsy and ridiculous.

The cockpit is near faultless, the seats are comfortable, the driving position excellent and the only incongruous note resulted from the use of rubber matting where carpeting would seem more appropriate.

Almost all the negative remarks made by our drivers about the Fiat resulted from its small engine and its comparatively modest power. For around-town traffic where the indirect gears can compensate in some degree, the Fiat is lively, responsive and good fun. Out on the highway, however, the engine whirs out a lot of revs, far more than any of our drivers felt comfortable with. With the top down, as one described it, you have the feeling that an engine is chasing you and having to work hard to keep up.

Other notes about the Fiat include agreement that it is the best riding of the four cars, that the shift linkage is less crisp than with the front-engine-rear-drive cars, that the front quarter windows are non-functional and that the hood latch is tucked away inside the lockable, drop-down glovebox.

The overall impression is that the Fiat tends to respond best to a light and delicate touch. As one of our young experts put it, "It's a girl's car, really." And he's right.

Triumph Spitfire

THE SPITFIRE is the middle-between sports car of our group; a newer design than the Sprite but not so new as the Fiat, more expensive than the Fiat and Sprite but not so costly as the Datsun, prettier than the Datsun or Sprite but not so pretty as the Fiat. It is basically traditional British sports car in design with the engine in the front and a non-synchro first in a 4-speed manual transmission, but it does have swing-axle independent rear suspension instead of the time-honored beam.

All our drivers agreed that the Spitfire had one major fault—the tendency for the body to jack and lift the inside rear wheel in a turn. This tendency is embarrassing and awkward and from outside the car it is frightening to watch as the outside front corner goes down and the rear wheels begin pointing at each other. From the driver's seat it is somewhat less disturbing as you feel the body twitch over so then what you have to do is feather the throttle, juggle the steering wheel to flatten it out again and then continue to balance yourself and the car on this tightrope through the bend. Empi (Box 1120, Riverside, Calif. 92502) has a low-cost camber compensator (Part No. HCCX-125) for only $24.95 and there's no $25 you can spend that will do more good for the Spitfire's handling. Some decambering also is possible and will be desirable in many instances.

The Spitfire rates as very good so far as driving position and comfort are concerned. The steering wheel is well placed, the seats are comfortable, the controls are where they should be and so on. The H-pattern of the 4-speed shift is slightly skewed but one quickly acclimates to this.

The Spitfire has fewer instruments than our other test cars and these are grouped in an attractive wooden panel in the middle of the dash, a location our drivers found awkward. There is a padded grab bar on the passenger's side but no lockable compartment in the cockpit.

The entire nose of the Spitfire swings forward to offer what must be an ultimate in engine accessibility. The accelerator action is somewhat notchy but the action of the other controls is good and the ultra-tight turning circle makes it possible to park the Spitfire in a very small space (one overenthusiastic driver found himself scrubbing the front tires too, because that turning circle is not achieved with correct steering geometry).

The straight-line acceleration of the Spitfire was the best of the four cars we tested and in top speed was only very slightly slower than the more powerful Datsun. Helping in this area is the fact that the Spitfire has a smaller frontal area and is aerodynamically more efficient than the Midget or Datsun. The ride of the Spitfire is also better than its live-axle competitors, especially on poorer road surfaces. In optional equipment, the Spitfire has a full assortment and is the only one of the four to offer overdrive.

In summing up, we would fault the Spitfire on its swing-axle rear suspension but praise it for its driving position, performance, comfortable ride, pleasant appearance and engine accessibility.

The Marvelous Big Austin-Healey

The story of a significant sports car

BY CYRIL POSTHUMUS

PHOTOS FROM THE AUTHOR'S COLLECTION

IF PEDIGREES WERE really that important, the big Austin-Healey could never have been the car it was. Its origins were humble, with an Iron Age engine left over from the Stone Age Austin A90; at first it was noisy, crude and cramped, yet fast and handleable in a he-man kind of way. It sold in thousands upon thousands, and when they levered in a smoother 6-cylinder engine and a 4-speed gearbox, sales mushroomed even more. They refined it, they raced it, they rallied it, and they broke records with it. It was incredibly tough and solid, yet handsome and amazingly cheap. In short, it was a lovable bastard.

The Big Healey came off the lines from 1952 through 1967, by which time the pundits had been predicting its doom for the past five years. But it never died of old age; it was killed by the U.S. Federal Safety Regulations, which came at a stage in the Healey's career when a major redesign to meet them was no longer worthwhile. So the handsome, long-nosed lorries met their end, and their begetter, Donald Healey, watched and waited, and finally got himself enmeshed with plans for their replacement. The story of that car, the Jensen-Healey, will be found on the pages just preceding.

Genesis

THE BIG Austin-Healey reigned as the best-selling medium-sized sports car for 15 years—longer than the complete life-span of the famous W.O. Bentley Bentley. At first, of course, it was just a Healey. I well remember the 1952 Earls Court Motor Show, when Pegaso of Spain sent one of their extra-exotic Saoutchik-bodied devices called, I believe, the Thrill. With V-8, 4-cam engine, De Dion type rear axle, and a price tag of about $30,000, it was a pretty comprehensive thrill at that, but it couldn't overshadow the late entry on the nearby Healey stand.

At that time the Donald Healey Motor Co. Ltd of the Cape, Warwick, made several models; the 2.4 Riley-engined Tickford saloon and Abbott coupe, 3-liter Alvis-engined sports convertible, and the export-only 3848-cc and 4138-cc 6-cylinder Nash-Healeys. Plans for a new model began early in 1951 when Donald Healey returned from a tour of the States with a formula for a car which would appeal to both home and export markets. He said, "I wanted to produce a very fast, everyday road car with genuine sporting characteristics, capable of 100 mph, which would also be excep-

Builder Donald Healey introduces the first 100 to R&T publisher John Bond.

JERRY CHESEBROUGH PHOTO

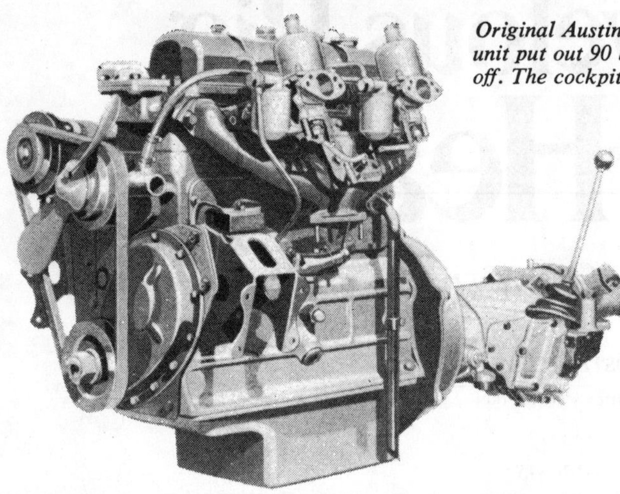

Original Austin-Healey 100 engine came from the Austin A90. This 2660-cc unit put out 90 bhp at 4000. Gearbox, also from A90, was 4-speed with 1st blocked off. The cockpit, while not exactly Spartan, had simplicity and elegance.

Austin-Healey History

tionally cheap to buy, and easy and economic to maintain." The answer was obviously a fairly lightweight and well-shaped car using mass production stock components wherever possible and the design team headed by Donald and his son Geoffrey set to work.

They evolved a simple chassis frame using two main 3-in. square box-section longerons, running parallel the length of the car, and some 17 inches apart. These were cross-braced by parallel and cruciform members, carrying steel floor pressings, and an integral cowl structure was built up from this base. The old Healey trailing link independent front suspension system was scrapped, being replaced by a coil and wishbone setup with lever arm shock dampers. Similar units controlled the leaf spring suspended live rear axle, and stock Austin parts figured wherever possible. These included the A90's 2660-cc 4-cylinder pushrod overhead valve engine, fitted with twin SU H4 carburetors, and producing 90 bhp at 4000 rpm. The standard 4-speed gearbox was also adopted, modified from column to floor shift, and A90 disc wheels were used.

This first Healey "Hundred" was finished in time for the 1952 London Show and rushed into place overnight. It was badly placed, tucked behind a pillar on the Healey stand but the public soon spotted the comely pale blue roadster with the ingenious slide-forward, tilt-back windscreen, and the crush soon became so great that protective barriers had to be put around the newcomer. The men from Warwick were flooded with orders, mostly from the U.S. where $7,000,000-worth of Hundreds were sold in double quick time. Leonard Lord and Lord Nuffield had been keenly interested in the new car ever since Donald Healey first contacted them about supplying parts, and now they stepped in with an offer to take on full production responsibilities, leaving the small Warwick company free for development and design. Overnight the deal was settled and the Hundred appeared at Earls Court next morning wearing new Austin-Healey wings on its nose. At the subsequent Miami World's Fair the new car won the Grand Premier Award, and at the New York Show it was voted International Show Car of the Year.

Road tests of the prototype showed that its power-to-weight ratio was so good that the very low first gear was unnecessary, so the production Austin-Healey Hundred or 100 was offered with first gear blanked off, producing a 3-speed all-synchromesh unit. Further tests showed that the engine would take an overdrive on the 4.1:1 rear axle ratio, and a Laycock de Normanville unit was fitted, operating on 2nd and top.

About 20 A-H 100s were built at Warwick to satisfy the dealers' initial clamor for something to show, and early tests proved the car was comfortably entitled to its official "100" enumeration. Donald Healey himself had in fact already exceeded 100 mph when John Bolster of *Autosport*

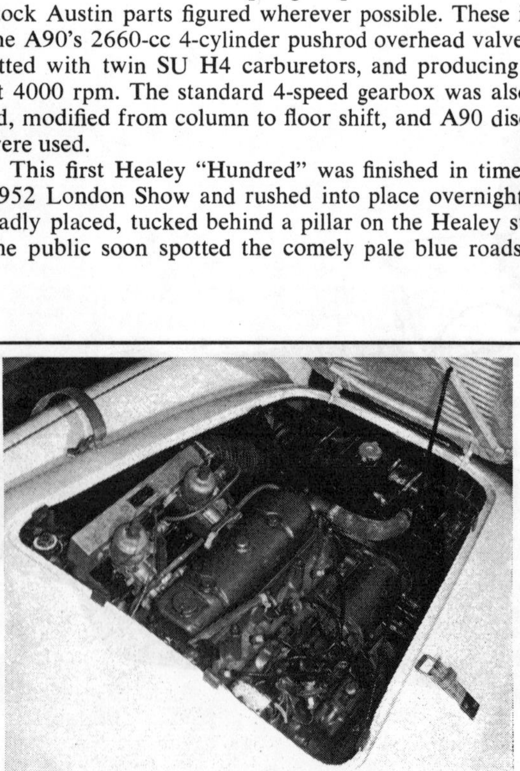

Austin-Healey 100M (for Modified) introduced in 1955 was customer's competition car and scored many successes in both European and U.S. club racing events.

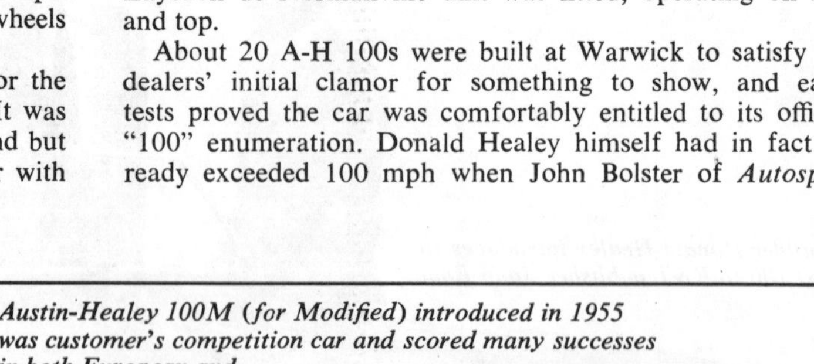

Austin-Healey 100-Six, introduced in 1956, had 6-cyl BMC engine from Austin A105 with 2639 cc and 102 bhp at 4600 rpm. Revised bodywork and top made it into a 2 plus 2. Wheelbase was increased 2 in., ahead of doors.

took the prototype to the Jabbeke-Aeltre autoroute near Ostend in Belgium, setting a 2-way average of 106.05 mph. Overall fuel consumption came out at 25 mpg, and it's nice to note that in my own contribution to *Autosport's* Motor Show copy I remarked ". . .it may well prove to be an important dollar earner. . ." Hrrrmph.

Series production began at Longbridge in May 1953, and the designation BN1 was bestowed upon the new model. In August 1955 the improved BN2 100 was introduced, using a 4-speed BMC C-type gearbox with overdrive, and with a hypoid rear axle replacing the earlier spiral bevel unit. Longer front coil springs were also fitted, along with improved brakes using wider shoes. The BN2 proved slightly faster on top gear and on acceleration with the improved transmission. Competition 100S and 100M models now came along, all using the 1949 2.6-liter engine, and 100 production over those four years totalled just over 15,000.

The Sixes

In September 1956 the new BN4 Austin Healey 100-Six was introduced, using the 6-cylinder BMC C-series 2639-cc engine from the Austin A105/Westminster, and featuring updated stying accommodation and fittings. The new 79.4 x 89-mm engine produced 102 bhp at 4600 rpm, no great improvement on the Four, particularly since the new car was considerably heavier. The point was that the Six's flexibility and smoothness made it a much more restful and easy car to drive, making the old Four seem lorry-like. The BN2's 4-speed gearbox was mated to the new engine, still with overdrive on 3rd and top, and other changes included enlarging the cockpit (at the expense of some trunk space) to include two very occasional extra seats. The wheelbase was increased slightly to make room for these changes, spring rates were altered to cope with weight distribution changes, and the 100's original folding screen was scrapped in favor of a fixed affair, plus a much improved hood and sliding side screens. Styling changes included the disappearance of the rather nice original Healey wedge radiator grille, which gave way to an oval grille as worn by the first competition 100S models. Rear deflectors were faired into the tail panels, an external fuel filler was added, and the bonnet line was generally lowered, acquiring a ventilation scoop to clear high spots of the new engine. Disc wheels became standard, but wire wheels remained optional.

Production of the BN4 commenced at Longbridge in August 1956, and moved down to Abingdon (originally MG's Pavlova works) in late 1957, when all BMC sports car production was centralized there. Just after the move, major mods were made to the Six's engine, which acquired a new 6-port head, an aluminum alloy inlet manifold, twin 1.75-inch SU HD6 carbureters, and a modified distributor. Compression went up from 8.25 to 8.5:1, and output rose to 117 bhp at 5000 rpm. It was well worthwhile; whereas the original BN4 100-Six accelerated from 0 to 60 mph in 12.9 sec and covered the standing ¼-mile in 18.8, the revised BN6 did 11.2 and 18.1 sec, respectively. Maximum speed in

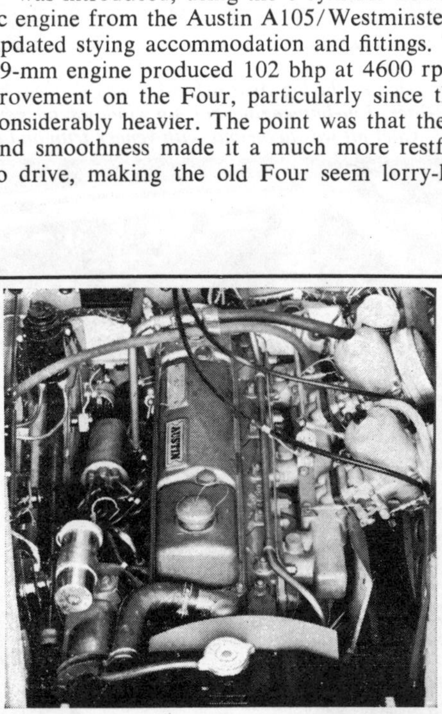

Popular demand also required that 100-Six be built as 2-seater, so this was revived in mid-1958 after introduction of 6-port engine.

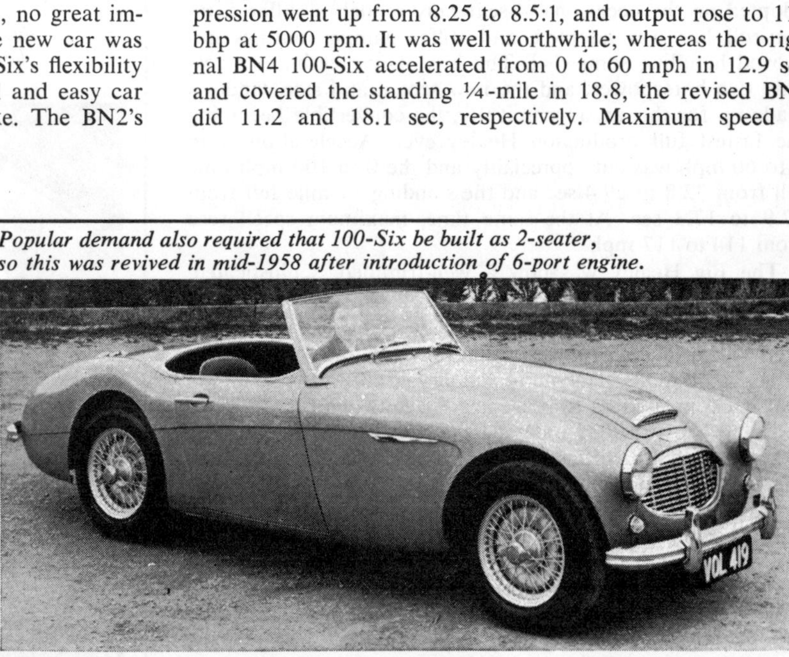

Austin-Healey History

overdrive top rose from 103 to 111 mph, and production of the 6-port BN6 ran on until March 1959. And though it had been specially designed as a 2/4 seater, there was still tremendous demand for a pure 2-seater, so they revived the latter from June 1958 on.

The 3-liter Cars

EXACTLY A YEAR later, BMC announced the enlarged C-series 6-cylinder engine, bored and slightly stroked out to dimensions of 83.3 x 88.9 mm (2912 cc). Compression was raised to 9.0:1, and twin SU HD6 carburetors were fitted with 124 bhp at 4600 rpm resulting. This unit went into the existing BN6 chassis/body shell with other modifications including a larger diameter clutch, stronger gear clusters and, following their success on competition 100-Six models, Girling disc brakes at the front. Both the 2-seater BN7 and 2/4 seater BT7 models were produced, with overdrive extra, and externally they scarcely differed from the earlier Sixes apart from a flash on the radiator grille and distinctive "3000" insignia.

In May 1961 a new mod inherited from competition was the fitting of triple 1.5-in. SU HS4 carburetors, which raised the output to 132 bhp at 4750 rpm, aided by a new camshaft and stronger valve springs. The radiator grille also received a convex vertical slat treatment to distinguish this, the 3000 Mk II, from the Mark I. Its extra horses mostly showed themselves in the upper reaches of the rev range; an added bonus was that the triple SU set-up, when properly tuned, was more economical than the original twins, but unfortunately such delicate tuning proved beyond the abilities of many private owners, and a return to twin carbs was made in March 1962. Popular optional extras then offered included servo-assisted brakes and a remote gearshift lever which allowed a fiberglass gearbox cover to be fitted, this in turn improving sound deadening and cockpit cooling, and lessening transmission noise.

At the time triple carbs were dropped, a restyled BJ7 3000 Convertible was announced, further refined, and slowly moving further away from the Big Healey's popular rorty image. A wrap-around windscreen, wind-up side windows, and a new one-piece detachable top with removable rear panel were adopted on the occasional 4-seat only model, a stiffer front anti-roll bar was adopted, and stiffer damper settings improved handling somewhat. The loss of one carbureter cost only two horsepower, and the new screen and side windows made up for this aerodynamically, the convertible becoming the fastest full production Healey ever. Acceleration from 0 to 60 mph was cut appreciably and the 0 to 100 mph time fell from 32.8 to 29.4 sec and the standing ¼-mile fell from 17.9 to 17.3 sec. At the same time, maximum speed rose from 114 to 117 mph.

The Big Healey was now a well-trimmed, sophisticated, and quite roomy GT, with plenty of smooth power and torque, still too little ground clearance, and a solid, heavy feel which made it quite a handful, if a pleasant one, to drive. More mods came up in January 1963; they altered the rear cockpit molding to prevent damage to the top on stowage, and in April they modified the hood assembly itself. In February 1964 a further refined Mark III variant appeared; they'd got the power up to 150 bhp at 5250 rpm, servo assisted braking was standardized, and a wood veneer facia with lockable glove box and a handy console built around the gear lever were other refinements. At this stage a new 3000 Sports Convertible cost £1185 on the UK market—ridiculously cheap for so much motor car, and comparing significantly with 1959 prices, when the BN7 2-seater cost £1257.

But the end was in sight. The U.S. Federal Safety Regulations were announced, coming into effect January 1, 1968. To build the necessary reinforcements into the Big Healey meant unwarranted structural alterations, the edict went forth from BMC to taper off production, and the 15-year run of this remarkable British sports car came to an end. Number 43,026, the last 3000 Mk III to be built, emerged at the end of 1967, bound for Jacksonville, Fla. In all, some 73,000 Big Healeys had been built since 1952, but now they had reached the end of the road.

The Big Healey in competition

DONALD HEALEY'S most successful sports car didn't sell on its performance/price/looks qualities alone; its ability had to be proved in competition, and proved it was, by every means that nimble publicity minds could think up. I have already related how the title Hundred was quickly justified by that 106 mph average at Jabbeke in 1952, but in fact, a few days before the car was announced, Donald Healey had taken one of the prototypes on to the Jabbeke road, recording a top speed of 113 mph, and picking up Belgian class records for the flying kilometer at 111.7 mph and for the flying mile at 110.9 mph. These early prototypes from Warwick were lighter than the Longbridge production models, but the 100's competition potential was obvious. The late Gregor Grant, founder/editor of *Autosport*, and Peter Reece from Liverpool gave the car its competition baptism in March 1953, entering the actual Earls Court car for the Lyons-Charbonnieres Rally, a tough 3-day affair over 1250 snow-packed miles in the mountains of Central France. The car went well until they hit a deep gully and broke a rear damper, eventually finishing with an almost detached rear axle and the springs munching the bodywork, in a lowly 50th place overall—after running 4th earlier.

This inauspicious debut was not to be typical of the Healey's career in races, rallies and record runs during the next 15 years—and more, for they were used long after production finally ceased and can still be seen in action today. At a time when 12-cylinder Ferraris, V-6 Lancias and XK120C Jaguars disputed sports car racing, works Healeys ran in the 1953 Mille Miglia (both retired) and at Le Mans (12th and 14th overall), and at the end of the year Donald Healey took

Final version of the big Austin-Healey was the Mark III. had 3-liter, 150-bhp engine, roll-up windows, convertible Cockpit was luxurious with veneer trim and glove

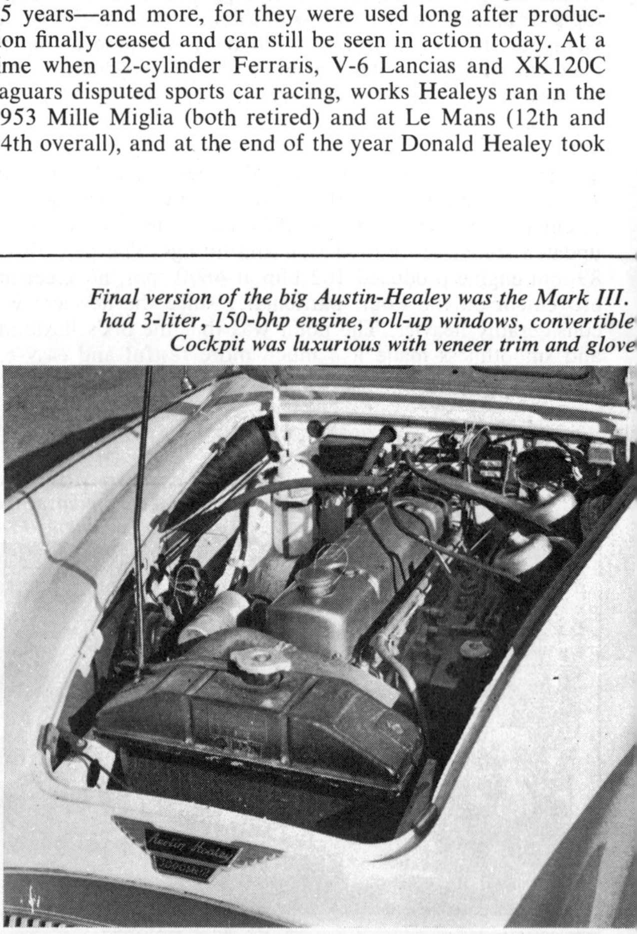

A total of more than 73,000 of the Big Healeys were built during the 15 years of its production.

a tuned model to the Bonneville salt flats, to attack International Class D records, and also drove a completely standard model, chosen by the AAA from a local dealer's stock, to attack U.S. stock car records. Driven by Healey himself, George Eyston, J.G. Benett, Roy Jackson-Moore and Jackie Cooper, the two cars between them captured over 100 records; the stock model set new U.S. figures in all categories from five to 3000 miles and averaged 104 mph over 24 hours, and the tuned 100 set new 12-hour figures and covered the measured mile at no less than 142.6 mph.

All of which immensely helped the image of "the world's cheapest 100 mph car," though the following year the Warwick works intensified their efforts in international motor racing. They developed a more powerful, lighter weight version of the BN2, to be sold in limited numbers and to form the basis of a tuning kit available to standard model owners. The works ran a prototype competition model at Sebring in 1954 with drivers Lance Macklin and George Huntoon. The car had center-lock disc wheels, Dunlop disc brakes all around, and a highly tuned 2.6-liter engine. It was hot for a Healey, but not in contrast to the costly front-liners. But as the Cunninghams, Aston Martins, Lancias and Ferraris dropped out, the Healey 100 climbed the leader board, amazingly taking 4th place at half-distance and getting to 3rd in the final hour. Out ahead were Moss with a brakeless Osca and a very tired 3rd-string works Lancia; Macklin made a final sprint for the lead when a sorely tried rocker arm broke and he had to nurse the car home on three cylinders, still a brilliant 3rd and class winner at 70.5 mph for the 12 hours.

After that, replicas of the 100S (S for Sebring) were in great demand, let alone the standard model, and the works' next venture was a 3-car team in the Mille Miglia for Macklin, Chiron and Tommy Wisdom (sharing with Morris-Goodall). This time things weren't so good; Macklin's was the first British car to finish, but only in 23rd place overall and 5th in class. International racing very much favored specialized racing machinery and Donald Healey felt so strongly about it that he issued a special statement and withdrew his team until changes were made in international regulations to give production machinery an even break.

So it was back to Bonneville in August, this time with the 1953 100 brought up to 100S specification, and also a new streamliner. The 100S's engine had an 8.3:1 compression ratio, twin SU H6 carbs and special manifolding, all resulting in 132 bhp at 4700 rpm; Dunlop disc brakes were used all around within wire wheels carrying Dunlop 5.50 x 15 racing tires, and the weight was down to 1870 lb. The stream-

The Racing Healeys

Above, one of four 100S models in the 1955 Mille Miglia. Below, Macklin/Huntoon 100S was 3rd overall in 1954 Sebring.

Big Healeys also rallied well: the Pat Moss/Ann Wisdom car that won Coupe des Dames and 2nd overall in 1960 Alpine Rally.

In 1954 the Healey Streamliner set 53 American and international speed records including a flying mile at 192.62 mph.

One of the less successful of the big Austin-Healeys was this endurance model fitted with long snout for Le Mans 24-hr race.

liner had a standard chassis and body with aerodynamic nose and tail attachments, a bubble cockpit canopy, and a fin behind the driver's head. Full 100S mods were made to the engine, plus the addition of a Shorrock supercharger, which sent the output soaring to 224 bhp. A 5-speed gearbox with built-in overdrive was used, and 16-in. disc wheels were fitted. Special safety provisions included an integral fire extinguisher system, 15 years before such things were featured in Formula 1, a self-jettisoning system for the canopy, and an automatic fuel cut-off operating on the oil pressure to prevent a serious blow-up.

The 100S prototype took the 12-hour record at 132.4 mph and the 24-hour at 132.29 mph, while in all Healey and co-driver Carroll Shelby took 53 American National and International 3-liter records. With the supercharged streamliner, the intrepid Donald Healey covered the flying mile at a resounding 192.62 mph! After which flush of success two cars ran in the Carrera Panamericana in Mexico, driven by Macklin/Healey and Shelby/Jackson-Moore. Shelby ran 4th on the opening stage, but then crashed, wrote the car off, and broke an arm, while Macklin went out early with electrical trouble.

An impressive season nonetheless, and before the 1954 London Show, Austin–Healey announced the competition 100S. The full specifications included a light alloy, 2-tone body with a smaller oval grille, louvered hood, Lucas Le Mans type headlights, cut down plastic windscreen, and a central filler cap feeding the 20-gal. fuel tank. No embellishers or bumpers were fitted, and the standard BN2 suspension was modified with stiffer springs and double-acting shocks. Dunlop disc brakes were standard all around, and overall weight was cut by some 220 lb.

The engine was much changed. A new light alloy Weslake cylinder head was used with the ports changed side-to-side; there were flat-topped, solid skirt high compression pistons, high lift camshaft, and combined oil cooler and filter mounted close beside the water radiator. Compression ratio went up to 8.3:1, and the 132 bhp at 4700 rpm was taken through a lightweight flywheel and competition clutch to the BN2 4-speed gearbox, shorn of its overdrive but with a modified remote change. The rear axle ratio was a surprisingly high 2.92:1, while many optional ratios were available.

The 100S accelerated from 0 to 60 mph in only 9.8 sec, covered the standing ¼-mile in 16.9 sec and had a top speed just on 125 mph. A total of 60 100S cars were built, and all save eight or nine were exported. A "Le Mans" engine conversion kit was offered to owners of standard BN1 and BN2 models who wanted some of the 100S performance even if they couldn't have the lightened shell. The kit comprised parts based on the 1953 Le Mans A-H 100 mods, offering twin 1.75-in. SUs with a cold air box, inlet manifolding, special distributor, high lift camshaft and 8.1:1 pistons.

Later these improvements were built into a limited production model, the 100M, made at Warwick and including a competition front anti-roll bar, special front shocks, louvered hood and 2-tone paint finish. It could be bought new as a works-modified BN2, or customers could buy the kit and have it fitted to their existing cars. The 100M was announced in October 1955, at the close of a season in which three 100Ss had placed 1-2-3 in class at Sebring, driven by Moss/Macklin, Brewster/Rutan and Cook/Rand, while George Abecassis had brilliantly won his class in one in the Mille Miglia, finishing 11th overall. These were the highlights of an otherwise black season, particularly for Lance Macklin. His car was hit by Levegh's Mercedes-Benz at Le Mans, with the awful results so well known, while at Dundrod in the TT he ran into the bank to avoid the multiple shunt which cost the lives of Jim Mayers and Bill Smith—disasters which blazed the Austin–Healey name uncomfortably across the pages of the sensational press. Nor was the 1956 season much compensation, for the A90 engine was now getting very long in the tooth, though the few 100Ss left in Britain did quite well in club events.

Then came the new C-series 100-Six, announced in September 1956. The release was preceded by another flurry of record breaking at Utah with C-series engined cars, Shelby and Jackson-Moore driving a BN2 with aerodynamic nose and tail sections, and fitted with the 150-bhp engine with 9:1 compression 6-port head and triple Webers; they took many records including the 500 miles at 153.14 mph and the 6-hour at 145.96 mph. Donald Healey drove the second car, one of the 1954 streamliners modified by Austin with a 250 bhp blown C-series engine. Unfortunately supercharger problems spoiled his attempts, but he did manage one good run timed at 203.06 mph, and great play was made of this figure in release publicity for the new 100-Six.

In March 1957 the type made its true competition debut. The BMC Competitions department had just been established at Abingdon and now the factory began to concentrate on international rallying. The debut was inauspicious (10th in class and 83rd overall in the Italian Sestrière Rally) but driver Tommy Wisdom, who had shared the rally car with his daughter Ann, redeemed himself in the last Mille Miglia by winning his class.

In 1958 several teams of 100-Sixes did well in several international rallies, winning both the manufacturer and team awards in the Liège-Rome-Liège with Pat Moss and Ann Wisdom taking the Coupe des Dames.

Although the 3-liter engine made its bow in 1959, the 100-Six pressed on in competition trim and in addition to rally work, a team of Cambridge undergraduates ran a 100-Six at Montlhéry to break seven international Class D records, including the four days and 10,000 miles at just over 97 mph.

The rally successes continued after the switch to the big 3000 in 1959 and this continued until BMC tapered off the Big Healey's competitive career in 1965 as emphasis shifted to the smaller, nimbler Mini-Coopers. At Sebring Paul Hawkins and Warwick Banks won their class that year.

In 1966 only privateers ran Big Healeys but when the 1967 RAC Rally included a class for Group 6 sports cars, Peter Browning, then competition manager, owned a 3000, ARX92B (re-registered as PWB 57) and this was built into probably the most powerful road-going Healey ever. An all-aluminum engine was used, bored out to 2968 cc, fitted with triple Weber 45DCOEs and producing just on 200 bhp at the rear wheels. But this last effort fell into anti-climax as the rally was cancelled because of a nation-wide epidemic of foot-and-mouth cattle disease.

Preparation

ALTOGETHER THE competition department at Abingdon operated 23 competition 3000s, plus one 100-Six brought up to 3000 specification. These were not specially built one-offs, but standard vehicles taken from the production lines and modified.

Between them, the 23 cars competed in more than 70 major international events, and most of the survivors have now been sold to private owners, several being regular club competitors in English production sports car racing and still picking up shiny pots almost every weekend. They're just as harsh and tough, heavy but responsive to drive as ever they were, but like old soldiers they just don't fade away . . . the big, beautiful bastards.

AUSTIN-HEALEY SPRITE/MG MIDGET, 1958-1970

Good handling and simplified design characterize these inexpensive sports cars

BY THOS L. BRYANT

FORMER R&T PUBLISHER John Bond once wrote, "To be successfully manufactured and sold, for minimum price, a car has to be of utmost simplicity in order to keep manufacturing costs low. Sports cars will never sell in the same quantity as family-type cars, so the argument that a sports car can sell as cheaply as a VW is not entirely valid" ("Miscellaneous Ramblings," July 1958). John was discussing the introduction of the new Austin-Healey Sprite, announced in May 1958 and destined to be as successful in its own way as the medium-size Austin-Healey 100 and 3000.

In the flush of success with the Big Healeys, Donald Healey decided to produce a smaller, more economical sporting car in the least expensive price range. As with the Big Healey, he based the Sprite on an existing sedan, the Austin A35, from which he took the 948-cc overhead valve 4-cylinder engine. This unit was slightly tuned for the Sprite with twin SU carburetors, special valve springs, improved exhaust valves and modified crankshaft bearings. This small displacement engine put out 48 bhp at 5000 rpm and 52 lb-ft of torque at 3300 rpm.

A 6-in. single dry-plate clutch connected the engine to a 4-speed manual gearbox with synchromesh on 2nd, 3rd and 4th. Ratios were 3.63 for 1st, 2.37 for 2nd, 1.41 for 3rd and 1.00:1 in 4th. The rear axle ratio was 4.22:1.

The rack-and-pinion steering system went from lock to lock in just 2.3 turns, giving the car very quick steering that took some getting used to initially. In our first road test of a Sprite (August 1958) we said, "In fact, the steering is nearly perfect for the purpose, and light and accurate besides."

The front suspension was also taken from the Austin A35 and consisted of lower A-arms and coil springs with lever-type hydraulic damper arms providing the upper suspension link. The solid rear axle was located by quarter-elliptic springs and lever-type shock absorbers. The Sprite came with 7-in. diameter drum brakes and 13 x 3½-in. steel wheels shod with 5.20-13 bias-ply tires at all four corners. This all worked quite well and cornering characteristics were close to neutral with a small amount of understeer at moderate speeds. It was very difficult to get the rear end to break loose and when it did it could be controlled quite easily.

While all of the drivetrain components and underpinnings were very conventional, the Sprite body was uniquely odd in appearance—so much so, in fact, that after awhile it began to look rather attractive! This apparent contradiction struck most of us in those early Sprite days as most of the motoring journals of the time had unkind remarks for the shape and design of the car. Soon, the bulging headlights and smiling grille were characterized as giving the Sprite a frog-like appearance and it became known as the frog-eye or Bugeye Sprite. Originally, Donald Healey's design had called for retractable headlights but when BMC actually began producing the car, this design was deemed too expensive and they were left perched in the middle of the hood.

The body of the car was a pressed steel shell and the entire

Austin-Healey Sebring Sprite.

Sprite dashboards weren't known for their over-instrumentation.

front end was hinged at the cowl to give easy access to the engine, steering and front suspension. The rest of the Sprite was fairly neat and clean with a simple, slightly sloping rear deck unmarred by a trunk lid. Entry to the trunk was from within the car and although the seatbacks were hinged to move forward, access to the luggage space was limited. John Bond described the appearance of the Sprite as a "hybrid of TR-3, Berkeley and Crosley Hotshot" and we are inclined to let it go at that.

The original Sprite was a classic roadster with removable soft top and side curtains. It was reasonably roomy inside with more pedal and leg room than the MGA but getting inside to make use of that room could be rather challenging, especially for a driver more than 5 ft 10 in. tall. The top came down far enough to restrict vision to the sides somewhat and rainy weather often pointed up a few places where the fit was not absolutely water tight, but then this was a sports car in the traditional sense of the word and comfort was of secondary or perhaps tertiary importance.

The Sprite carved a new niche for itself in the sports car world, offering superior and economical performance for a price comparable to the least expensive sedans of the day. Almost immediately, Sprites began to appear at sports car races all over the world and soon dominated H production racing in the U.S. Who could ever forget those days of the late Fifties and early Sixties when a gaggle of the little cars would take off sounding for all the world like a horde of angry bees chasing a hive-robbing bear! Not only did club racers turn to the Sprite; such big names as Stirling Moss, Walt Hangsen, Bruce McLaren, Briggs Cunningham and others competed in works cars or the Sebring Sprites of John Sprinzel.

Sprite Mark II

THREE YEARS after the introduction of the original Sprite and some 49,000 cars later, the Mark II version was unveiled in May 1961. Gone were the frog eyes, replaced by conven-

Austin-Healey Sprite Mark II.

tionally located headlights in fixed fenders, along with a wider grille and a hood that opened without lifting the entire front body just like most other cars. The rear was also restyled with a trunk lid that opened from outside and a 12-in. cut in the back of the cockpit for luggage or a small child.

There were also changes to the engine, including an increase in the compression ratio from 8.3 to 9.0:1, larger throats in the SU carburetors, larger intake valves and a change in the exhaust valve timing. The result was more power (50 vs 48 bhp) without loss of torque. The gearbox was also revised; the close-ratio gears used on the Sebring Sprites in competition and available as an option for a year or so were now standard in the new car.

To the avid Bugeye Sprite fanatic, the Mark II was an abomination. To the rest of the motoring enthusiast world, however, the new car was a definite improvement: "Of course, the new model may be accused of some minor loss of personality, but no one can deny that the Sprite II is better looking" (R&T, August 1961). The Mark II was also a more convenient car to live with because of its outside opening trunk and better use of space. R&T concluded that the Mark II, like the original Sprite, "offers more fun per dollar than anything we have driven for a long time."

Only a month after the announcement of the Sprite II, the same car was brought out under the MG banner as the Midget. The only differences were in the nameplates and a few minor trim details and this practice was to continue until the eventual demise of the Sprite name in 1970, when the Midget continued on by itself.

Sprite 1100

Four-and-a-half years after the introduction of the Sprite, a new 1100 model was shown at the London Motor Show in October 1962. The new Sprite (and Midget) had been upgraded with an increase in engine displacement from 948 to 1098 cc, which raised the horsepower rating from 50 to 55 at 5500 rpm. The clutch diameter was increased by 1 in., front brakes were now disc rather than drum, and reshaped and more thickly padded seats plus the addition of carpeting helped to make the Sprite more comfortable.

The additional engine displacement was the result of an increase in bore (2.54 vs 2.48 in.) and a new, longer-stroke crankshaft very similar to the one used in the MG 1100 sedans. The displacement increase made the Sprite more pleasant to drive because of the extra torque available (61 lb-ft at 2500 rpm compared with 52.5 at 2750 in the Mark II).

Although the new seats were certainly more comfortable and the noise level had been reduced by the carpeting, the Sprite was still characterized by many of its original features: no wind-up windows, separate key and starter (the latter a pull cable rather than button), no door locks or outside door handles and no glovebox. The only lockable portion of the entire car was the trunk.

In terms of performance, the Sprite 1100 was a much better car. Top speed improved by only a few mph but the change in the torque characteristics offered top gear hill climbing and easier passing than in previous Sprites. This new found performance could be used with increased confidence as a result of the change to front disc brakes. The former drum brakes were marginal when the car was being driven near the limit. Handling was unaffected by any of these changes and the Sprite (and Midget) was still one of the most responsive cars on the road.

Perhaps more amazing than anything else was the fact that the new Sprite buyer was still getting all of this for less than $2000! In the period from May 1958 to October 1962, the Sprite's list price had only gone from $1795 to $1985. And *Road & Track* was still saying that this car offered more fun per dollar than any other, even after 4½ years.

Sprite Mark III

March 1964 brought the introduction of the Mark III Sprite (Mark II Midget which was one number behind) and

this new model reflected the need to bring the Sprite up to date with the modern sports car. Major improvements included wind-up windows, swiveling vent wings, updated instrument panel and further improvements in interior trim.

The rear suspension came in for revision with a change from the quarter-elliptic springs to semi-elliptics to reduce the tricky roll oversteer inherent in the original design. There was also a minor improvement in engine performance (stepped up to 59 bhp) through improved manifolding and the use of the MG 1100 cylinder head.

Sprite Mark IV/Midget Mark III

Once again the London Motor Show was the arena for the display of the new Sprite/Midget, this time in 1966. The car now became a thoroughly modern sports car with a proper

convertible top that could be raised and lowered without dismantling. Another engine transplant had taken place and the Sprite/Midget now had a 1275-cc powerplant similar to that used in the Mini Cooper S but detuned from 75 bhp to 65. This detuning allowed lower production costs because a normal forged crankshaft could be used in place of the more expensive nitrided steel crank of the S for example, while still maintaining the Sprite's reputation for reliability and long life. The net increase of 6 bhp over the Mark III Sprite (Mark II Midget) was enough to make a surprising improvement in performance as the new Sprite/Midget would accelerate to 60 mph in 14.7 seconds versus 18.3 sec for the previous model.

The crisp handling which had always been a characteristic

of Sprites remained and there was still some roll oversteer built into the rear suspension that made the car great fun to drive sideways. The ride was still a bit jouncy over uneven surfaces but it was truly sporting in nature and aficionados of the breed didn't mind it at all and in fact felt (and still feel!) that it was necessary or you might as well have been driving a large sedan.

The Sprite/Midget cars remained little changed from this setup until 1975. Of course, the Sprite version was discontinued in 1970. The Midget continued without any drastic changes except for the addition of more and more emission controls and safety items such as the over-large bumpers of today, until the 1975 model which received the 1493-cc Triumph Spitfire engine with slightly different exhaust manifolding. Emission controls had become stringent enough that the increased displacement was capable of putting out 55.5 bhp at 5000 rpm.

For the purposes of our discussion here, we have elected to cut off our report with 1970 for two reasons: that was the last year for both the Sprite and the Midget; and the increasing emission and safety regulations have taken some of the fun out of driving the more modern Midgets. They still have reasonable handling, are relatively lightweight and thrifty of fuel, but in talking about a Used Car Classic we also have to give consideration to purchase price and post-1970 car prices are generally out of the bargain class.

Car selection tips

As WITH most British sports cars of the time, the early Sprite/Midget is a remarkably sturdy and simple car. The engine and transmission having come from a sedan of some years' standing, they were time-tested and of proven reliability. The twin SU carburetors have a tendency to be touchy in adjustment so many times a car that does not seem to be working properly may just need a delicate hand applied to the carburetors. Also, the linkage runs right into the carburetor throat and there have been cases where the opening becomes worn, allowing air to seep in and upset the mixture. The use of some rubber grommets can cure this malady. One of the few weak spots in the engine is the center main bearing on the crankshaft. If it's at all suspect, replace it.

All of the Sprite/Midgets covered in our time period have a non-synchromesh 1st gear. While the gearbox is sturdy, a heavy-handed driver can wreak damage on the unit, especially 1st, and it may be necessary to consider rebuilding it. On the other hand, we have been told by Sprite owners that they have put considerably more than 100,000 miles on their cars without the gearbox showing signs of wear.

Rust is one of the most important things to look for in buying any car and it's true of the Sprite too. The potential buyer may be fortunate and find a car that has a number of holes drilled in the rocker panels and the bottom edge of the hood (Bugeye models)—these are to permit water to drain out and prevent rusting.

Prices for used models vary from area to area of the country. The Bugeye is rapidly approaching classic status and the prices are beginning to reflect this. However, a running Bugeye that needs restoration can occasionally be found in the $600-800 range in southern California, but the usual price is $200-400 more than that. The fully restored Bugeye can bring as much as $2000 and more. Prices will generally be somewhat higher

TYPICAL ASKING PRICES

Year & Type	Price Range
1958–61 Austin-Healey Sprite	$600–1200
1961–62 Austin-Healey Sprite II & MG Midget	$500–1200
1963 Austin-Healey Sprite 1100 & MG Midget 1100	$400–1200
1964–65 Austin-Healey Sprite III & MG Midget II	$500–1500
1966–70 Austin-Healey Sprite IV & MG Midget III	$800–1800

PERFORMANCE DATA
From Contemporary Tests

	1958 Austin-Healey Sprite	1961 Austin-Healey Sprite II	1963 Austin-Healey Sprite 1100	1967 MG Midget III
0–60 mph, sec	20.8	19.6	18.3	14.7
0–80 mph, sec	35.5*	49.0	42.5	31.0
Standing ¼ mi, sec	21.8	21.5	20.9	19.9
Avg fuel economy, mpg	34.0	34.0	33.0	24.0
Road test date	8-58	8-61	8-63	9-67

*0–70 mph

BRIEF SPECIFICATIONS

	1958 Austin-Healey Sprite	1961 Austin-Healey Sprite II	1963 Austin-Healey Sprite 1100	1967 MG Midget III
Curb weight, lb	1460	1540	1560	1560
Wheelbase, in.	80.0	80.0	80.0	80.0
Track, f/r	45.4	45.8	47.2	46.3
	44.8	44.8	45.0	44.8
Length	137.0	136.0	138.0	137.4
Width	54.0	54.0	54.0	56.5
Height	48.0	48.8	47.8	48.6
Fuel capacity, gal.	6.0	7.2	7.2	7.5
Engine type	ohv inline 4	ohv inline 4	ohv inline 4	ohv inline 4
Bore x stroke, mm	63.0 x 76.2	63.0 x 76.2	64.5 x 83.8	70.6 x 81.3
Displacement, cc	948	948	1098	1275
Compression ratio	8.3:1	9.0:1	8.9:1	8.8:1
Bhp @ rpm, SAE gross	48 @ 5000	50 @ 5500	55 @ 5500	65 @ 6000
Torque @ rpm	52 @ 3300	52.5 @ 2750	61 @ 2500	72 @ 3000
Gearbox	all models: 4-speed, non-synchro 1st gear			
Final drive ratio	4.22:1	4.22:1	4.22:1	4.22:1

in other parts of the country where these particular models are less readily available. Prices on newer Sprites and the MG Midgets will run anywhere from $800 on up, peaking at about $1800.

The Sprite/Midget series of cars is perhaps one of the most logical ones for inclusion in our Used Car Classic series of reports. They are simple, relatively easy to maintain, and, just as when they were new, return more driving fun per dollar than just about any sports car we can think of. If you are the proper size to fit inside one, you couldn't do much better.

DRIVING IMPRESSIONS

PHOTO BY DOROTHY CLENDENIN

ONCE AGAIN WE find ourselves indebted to Ken Schwartz of the Long Beach (California) MG Club for leading us to a Used Car Classic owner who was willing to have us photograph and drive his car. Eddie Martinez runs a speedometer repair shop in Long Beach, owns two Bugeye Sprites and writes music and performs in musical productions. He is one of those rare persons who seems to enjoy life just a little more than the rest of us and his Sprites provide him with a lot of enjoyment—and frustration, he adds.

Eddie has been a sprite owner for 12 years and both of his Bugeyes are of 1959 vintage, both are painted bright yellow and both have received his personal attention in interior improvements. These include re-doing the dash, adding a center console and installing a vinyl-covered plywood bulkhead between the seats and the trunk. Eddie claims this last item has cut down on the noise level considerably and he is justifiably proud of his work.

To the accompanying snide comments and laughter of Eddie and R&T Managing Editor Dottie Clendenin who came along as the official photographer, I gingerly eased my portly 6 ft 2 in. bulk into Eddie's roadster. His second car had the hardtop on it and I wasn't even about to try getting into that one! Gad, but these are tiny little devils. I soon found that I was indeed in (no need for a seatbelt here as it would take more than a collision to pry me out) and that I could operate the pedals. I could not however put my left foot anyplace once I had activated the clutch and let it up, so I drove along holding it in the air just above the pedal. The controls fell readily to hand--there was no place else for them to fall—and with a turn of the key and a pull on the starter we were off.

Eddie mentioned to me as we were motoring along that I looked sort of funny in the car—I suppose he meant that I bore a strange resemblance to a trained circus bear driving a kiddy car—but I was undaunted, dividing my time between bending down to look through the windshield and craning upward to look over it. The day was one of those magnificent California days after a rainstorm, with a freshness to the air we rarely enjoy. The wind was in my face and the exhaust was singing with a healthy note. "This is what it's all about," I told myself as my left leg began to cramp up from being suspended in air for half a mile.

The ride was sheer sports car delight—jouncy and stiff, giving the driver and passenger intimate knowledge of each and every surface irregularity along the way. At the same time, however, there is a feeling of maneuverability and control so that the bumps fade away, replaced by a fiendish desire to cut and slice through and around all those behemoths blocking the way. The 948-cc inline 4-cylinder engine revs freely up to about 4000 rpm and then begins to hint at a bit of strain. Eddie said that the crankshaft in his car was beginning to make trouble, so we didn't run the engine up to the 6000-rpm limit. The handling is close to neutral and getting the tail to hang out takes a definite effort. Once you do, it's very easily controlled—in fact, you really have to work at it to get in over your head driving a Sprite, which is why they have always been so outstanding for the young sports car enthusiast learning the ropes.

The Bugeye has generally been considered an unappealing design but I've never agreed with that—it's strange enough to have a beauty all its own. The smiling front end, which teenage girls tend to label "cute," is actually a grin with a touch of a leer in it. And that's precisely what driving the Sprite is like. It's not fast, not terribly comfortable, not very refined. So why are all those Sprite/Midget owners grinning?

—Thos L. Bryant

OWNER IMPRESSIONS

MIKE GRISWOLD WAS getting married so I bought the Bugeye Sprite from him for $400. Then Bill Holley and I nursed the car from the wedding in Salt Lake City back to Madison, Wisconsin, sleeping under the stars and occasionally in the rain, eating cheap greaseburgers and only making it home because buried behind a pile of recap carcasses at the Goodyear store in Cody, Wyoming was one dusty 5.20-13, which they sold to me for cost.

Privation—it was a fitting introduction to my Sprite. That car was an adventure from the first time Holley hit a pot hole in Utah and we realized the right front shock was broken. That car taught me to be a very good diagnostic mechanic.

Yet, I've never again gotten as much pure automotive joy as I did from the Blue Bugeye. It was just me, my swimming trunks and this car with the top and side curtains I could remove and stow behind the seats in under 30 sec. The early morning runs in the cool, damp Wisconsin mornings with the wind knotting my hair and fighting to get inside my nylon ski jacket are still very alive. It was a summer of cut-off Levis, cut-off sleeve sweat shirts, white tennis shoes (no socks) and driving this little British roller skate to Elkhart Lake, Meadowdale or the USAC races at Milwaukee. The little bugger even crapped out one night so I could forsake it for a blind date and meet the lady who is now my wife.

Every reaction of that roadster was crisp and immediate—a sudden sneeze could prompt a lane change. Best of all, the car was joyfully simple, which is why they ruined it when they added roll-up windows and a fold-down top. Complication was the very antithesis of the Sprite. Why must we reject simplicity and call its complicated replacement "sophisticated?" The British, of all people, should know better. I still look at Bugeyes lovingly and occasionally think of buying another, this time in British Racing Green, with wire wheels and a Nardi steering wheel. *—John Lamm*

AUSTIN-HEALEYS, 1953–1967

Marvelous cars that changed the sports car world

BY THOS L. BRYANT
PHOTOS BY WM. A. MOTTA

ONE COULD WRITE reams about the history, charm and charisma of the Austin-Healey cars and their impact on the motoring enthusiast world. The Big Healeys (to distinguish them from the Sprite) were produced for some 15 years and carved out a niche for themselves as the best selling medium-size sports cars of their day. The Healey was something of a natural progression for the American sports car driver of the Fifties, who had started with an MG TC or TD and wanted to move up to a more powerful car. I remember this was the case with one of my brothers who owned a bright red 1952 MG TD, which he rather generously shared with me and in which I received my first traffic citation. Eventually the red TD gave way to a jet black Austin-Healey 100-4, which seemed to our youthful eyes ever so much more rakish and befitting our dashing self images. Unfortunately, we weren't wise enough to keep both cars and the Healey later went away in favor of something else. A DKW, I think.

The point is, the Austin-Healey was an important factor in the development of the sports car movement in America as well as in other parts of the world, and it's a car that still has a surprising number of devotees eight years after its demise. R&T Contributing Editor Cyril Posthumus wrote about the Big Healeys in April 1972. "Incredibly rough and solid, yet handsome and amazingly cheap. In short, it was a lovable bastard." Posthumus was referring to the fact that the original Healeys had humble origins, using the Austin A90 4-cylinder engine mated to a 4-speed gearbox that had such a low 1st gear it was blocked off in the earliest cars.

The Austin-Healey 100 made its debut at the London Automobile Show in 1952 and was an immediate hit. Donald Healey and his son, Geoffrey, designed the car to be a lightweight, well shaped automobile, using as many stock Austin components as possible to keep the cost down. The original A90 Austin engine was a 2660-cc overhead-valve affair which produced 90 bhp at 4000 rpm. Coupled with the A90 4-speed gearbox which had been slightly modified so that it could be floor-mounted rather than column shift, the combination gave the Healey an excellent power-to-weight relationship. The final drive ratio was 4.10:1 which further aided the acceleration but did little for high-speed cruising. Thus, a Laycock de Normanville overdrive unit was fitted.

The chassis was a relatively simple one with two main 3-in. square box-section frame rails with crossmembers that carried steel floor pressings. The front suspension consisted of coil springs and A-arms with lever-type shock absorbers. The rear suspension consisted of a live axle suspended and located by leaf springs and lever shocks.

The 100-4 went into production at Longbridge in May 1953 and continued little changed until June 1955. In August of that year, improvements were made to the brakes and front springs, and the 4-speed BMC C-type gearbox was installed. Almost 15,000 100-4s were built between the startup and November 1956, including the relatively rare 100M (for Le Mans) and very rare 100S (for Sebring) competition models.

The Sixes

IN THE late summer of 1956, the Austin-Healey 100-6 was born with the BMC C-series 2639-cc engine. This new powerplant produced 102 bhp at 4600 rpm, but this improvement in horsepower was pretty well eaten up by the increased weight of the new Healey. The cockpit had been enlarged at the expense of the trunk and two very small seats had been placed behind the front seats. There were minor revisions to the body, the wheelbase was slightly longer and the original fold-down windshield was replaced with a fixed one. The gearbox remained unchanged from the late 100-4s.

In 1957, modifications were made to the 6-cylinder engine in the form of a new 6-port head, aluminum alloy intake manifold, modified distributor and twin 1.75-in. SU HD6 carburetors. These changes brought the compression ratio up to 8.5:1 and the horsepower rose to 117 at 5000 rpm, giving the 100-6 a slightly quicker elapsed time for the quarter-mile run (18.1 vs 18.8 sec) and improved the top speed in overdrive from 103 to 111 mph.

While the bulk of 100-6 Healeys were of the new 4-seat configuration, the demand for a 2-seater was so strong that in June 1958 the factory began turning out 100-6s with only two seats. Again, almost 15,000 100-6 cars were produced, with more than 10,000 of them being 4-seaters. Production of the 100-6 ended in March 1959 and three months later the first of the 3000s emerged from the factory, which had by now been moved to Abingdon.

The Healey 3000

MORE THAN 42,000 Austin-Healey 3000s were built during the period from June 1959 to December 1967 when the famed marque came to an end. The 6-cylinder C-series engine was bored and stroked to 2912 cc, compression was raised to 9.0:1 and power went up to 124 bhp at 4600 rpm. The 3000 also featured a larger-diameter clutch, stronger cluster gears and the addition of Girling disc brakes. In appearance, the early 3000 was little changed from the 100-6, except for the flash on the grille and the 3000 insignia.

The 3000 Mark II was introduced in May 1961 and had three SU carburetors, a new camshaft and stronger valve springs, bringing the bhp figure up to 132 at 4750 rpm. The new triple-carb arrangement provided more power, mostly in the top end, and slightly better fuel economy when properly tuned, and those last three words are important. Difficulties in maintaining the proper state of tune were so widespread that the 3-carburetor setup lasted less than a year, and in March 1962 a return was made to twin carburetors.

Several changes were made to the body. These included a fully convertible top (as opposed to a roadster with removable top), a wrap-around windshield, wind-up side windows in place of curtains and other refinements that moved the Healey away from its traditional sporty image and toward a more comfortable touring car position. Performance did not deteriorate with the loss of the third carburetor and, in fact, the convertible became the fastest production Healey ever with a top speed of 117 mph and a quarter-mile time of 17.1 sec.

Over the next five years, until the end of production in December 1967, the Big Healeys were marked by modifications designed to make them well fitted, pleasant GT cars. The horsepower climbed to 150 at 5250 rpm, servo assist was added to the brake system and the interior was groomed with a wood veneer facia and lockable glovebox. The Austin-Healey was still marked by an abundance of smooth power and torque, along with a heavy feel to the handling that inspired confidence in the car's durability. The styling was timeless, one of the cleanest and best-looking sports cars ever. But the U.S. federal safety regulations were announced for January 1, 1968 and the factory decided that too many modifications would have to be made to meet them. After 15 years and more than 70,000 cars, Austin-Healey became a marque of the past.

Buying a Used Healey

THE PURPOSE of the R&T Used Car Classic series is to give the reader sufficient information to make a decision about purchasing the subject car. To many of us, buying a new sports or GT car may be financially difficult (if not impossible) or we may simply be caught up in a euphoria of nostalgia that tells us they just don't build them like they used to. Whatever the reason, great joy and exhilaration can result from owning a car such as the Austin-Healey, or the other marques we have covered in this series. It's also prudent to point out that equally great frustration and despair can arise if the buyer makes an unfortunate choice.

For this portion of the report, we solicited the advice of David Ramstad of Everett, Washington, a Healey owner and active

member of the Austin-Healey Club, Pacific Centre. Dave writes:

Thinking about adding an Austin-Healey to your stable? But hesitating because well meaning though ignorant friends have done their best to divert you from the heartbreak of cranky old English sports cars? Did they hit you with dire warnings about those devilishly complicated SU carburetors or could it have been hideous word pictures of wire wheels flying off worn splines at speed? While some British sports cars may have earned such reputations (I love them all, regardless), I don't mind shouting that the Austin-Healey is not among them! Now, let's get into the pleasant task of making a wise purchase. All the standard guidelines for choosing any used car apply and will not be dealt with at length here. Our main concern is things peculiar to the Austin-Healey.

Mechanical

THE POWERPLANT, be it A90 4-cylinder or C-series 6, is a rugged, low-stressed, heavy cast-iron unit producing a considerable amount of torque at low speeds. These engines are noted for their reliability and long lives, quite lacking in major trauma. Unlike the highly tuned engines found in Alfas, Porsches, etc., the Healey's engine almost never breaks, but simply increases its clearances with a corresponding increase in oil consumption over a comparatively long period of time. The engine is very easy to live with because of its ample power and infrequent mechanical needs and, happily, pollution control devices are absent because all Healeys were produced prior to 1968.

The prospective buyer should perform a compression check and if possible a full leak down test to see if copious amounts of oil are coming from the lower rear of the engine. If so, this indicates either a failure in the rear plate gasket or a crack in the rear main bearing area, among other less likely possibilities. If you are not planning to tear down the engine in your restoration process, you should avoid a car with this problem. The average, reasonably well cared for engine usually requires little more than a careful valve grind, piston ring replacement and complete tune-up to put it back in the pink. With good lubrication practices, the lower ends of these engines seem to last forever.

The BMC C-type 4-speed gearbox fitted to all Big Healeys from August 1955 on can be a problem if found in a car once owned by an insensitive (read brutal) gear changer. First and 2nd gears normally produce a distinct but not excessive whine, but the mauled gearbox will make obviously expensive noises (most often in 1st and reverse) and will usually exhibit slow or nonexistent synchromesh action. Because of the Healey's ample torque, synchromesh was never felt to be necessary in bottom gear.

The Laycock de Normanville overdrive unit seldom suffers failure in its internals. Poor or no overdrive function is often cured by careful troubleshooting of the unit's electrical control system or by adjustments to the hydraulics. If questions remain after a test drive, consult local Healey enthusiasts, British car repair shops or the nearest Leyland dealer, in that order.

The Big Healey's rear axle is very strong and trouble free, seldom needing more than a seal replacement to cure a leak. This axle is often called upon to handle the torque of an American V-8 and apparently suffices.

Body and Chassis

ENEMY NUMBER one to the A-H semi-unit, aluminum and steel construction is rust. The buyer should carefully examine the lower 12 in. all around the car to determine the extent of deterioration. Lower fender to chassis joints (front and rear), rear fender lower cavity, lower door area and sills and lower flange of the trunk lid are the most common areas for rust. Cars in the north central and eastern U.S. and eastern Canada, where salt is used on winter roads, have usually suffered the most. Many Healeys show some minor, non-structural rust which can be easily repaired, but those cars with major rust in the inner body sills, outrigger beams and floor sections are to be avoided.

I should stress the Big Healey can be kept free of rust with some precautions. The car has few totally closed areas and if water drains in doors, sills and convertible top channels (1963–1967 cars) are kept open, and if accumulations of road dirt and salt are frequently flushed out of wheel wells and lower body areas, the Healey can be preserved intact for many years. Speaking of rust problems, a fast-moving item in the Healey world is the replacement fuel tank. The tank is in the trunk cavity and will occasionally have deteriorated to the point of seepage. Fortunately, replacements are available for approximately $110.

The substantial box-section chassis appears to be indestructible. This is not an illusion—terribly bent Healeys have been reconstructed because of the strength of the frame. However, carefully examine the forward and aft crossmembers. Lack of care when jacking up the car at these points (Healeys are never raised by their bumpers!) has led to serious bending or holes punched through the welded seam by a small but lethal jack pad. Using jacks with wide flat pads or a block of wood will prevent damage of this sort.

Healey front suspension and steering may exhibit considerable looseness. The cure here is to overhaul and re-bush the lower A-arm pivot points and swivel pins, replace the tie rod ends and adjust or overhaul the steering box. You should recognize that the Healey's cam-and-peg steering lacks the taut feel of rack and pinion, and Healeys always display a certain amount of play in the steering wheel.

Rear leaf springs occasionally turn up with broken leaves, in which case the car's attitude will take on a marked list. The four Armstrong lever-type shock absorbers will need to be replaced if leaking badly. Americans haven't seen a car in years that requires refilling the shocks; often a neglected Healey may simply be in need of shock fluid to restore its proper ride and handling. The price of replacement lever shocks has risen to the point where the cost of converting to tube-type shocks is justified. This requires the addition of mounting brackets to the chassis, and these are available along with the proper shocks from major Koni suppliers.

The Achilles heel of most wire wheel-equipped sports cars is the splined hub on which the rear wheels are mounted and, of course, the wire wheels themselves. If you test drive a car that exhibits a momentary lag followed by a clunk from the rear when you engage the clutch, you face a large cash outlay to replace or re-machine the splined rear hubs. There has been a love-hate relationship with wire wheels forever and their periodic maintenance is costly. This is not helped by the fact that the noble profession of wheelwright is disappearing, and

Six-cylinder engine in early 3000 (foreground) displaced 2912 cc while the original 4-cyl engine (rear) was a 2660-cc affair with 90 bhp.

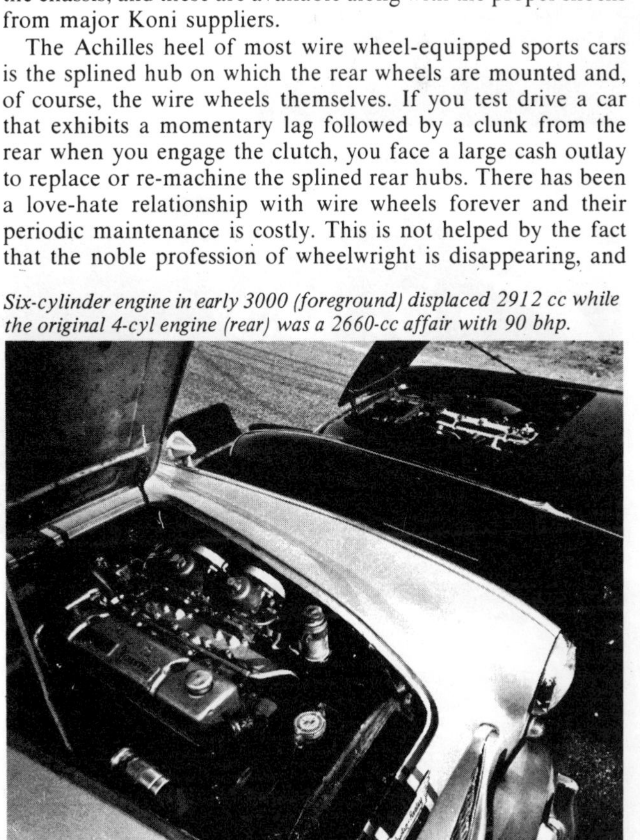

many owners do convert to stamped steel discs or cast alloy wheels. Few owners of traditional British sports cars can tolerate the accompanying character alteration, however. Replacement Dunlop wire wheels have increased drastically in price and you should plan on paying $70–$80 each for new wire wheels—possibly half that for rebuilding your existing wheels. Talk around before committing your wobbly rims to the local wheel wizard because quality varies considerably.

Parts and Prices

WHAT'S THE story concerning parts? Initially, as established dealer spares inventories shifted away from the Healey when it was discontinued, the situation looked quite bleak. Fortunately, circumstances began to turn around just a few years ago and today there is no need to fear any serious shortage of general spare parts. With the Big Healey's emergence as a collectible car there has been a rise in small, specialized businesses in North America and Great Britain dealing in used, new-old stock or newly manufactured copies of obsolete parts. Simply combing the "Market Place" section of *Road & Track* will yield several firms offering extensive lists of spares. You must remember though, that Healeys are not Beetles or Impalas so you may have to do a certain amount of legwork. Considering the car has been out of production for eight years and compared to other sporting cars of recent decades, parts are surprisingly reasonable and not terribly expensive.

How much is that solid, well maintained old Healey going to cost? While wild price fluctuations aren't uncommon in a car of this sort, some approximations can be made. There are very few original or acceptably restored 4-cylinder 100s (1953–1956) below $2000. I believe the same can be said for the 100-6 series (1957–1959) and the early 3000, unofficially known as the Mark I (1960–1961). The 3000 Mark II, the triple-carburetor equipped 1962 roadster and rollup window 1963 convertible, is most often found in the $2000–3000 range. And the final Big Healey, the 3000 Mark III (late 1964–1967), loaded with 150-bhp engine, vacuum-boosted disc brakes, and walnut facia is a bargain between $3000–4000, with figures of $4500–5000 not unusual. Extremes range from badly rusted machines going for a few hundred bucks in eastern Canada to pristine, low-mileage Mark IIIs listed at more than $6000 here in the Seattle area.

Conclusion

UNDAUNTED BY those who consider seekers of old British sports cars completely mad and equipped with the basic data necessary to distinguish a good prospect from a dog, the search for that fine old Austin-Healey can begin. At this point, a mighty plug for Healey clubs is in order. Far from being simply mutual admiration societies, these groups of true believers diligently pursue sources of spare parts, offer discounts to their members on various Healey-related merchandise and publish priceless technical and historical information. And that only accounts for half of their efforts. The number of club tours,

Dash of 100-4 is clean and simple.

BRIEF SPECIFICATIONS

	100-4	100-6	3000	Mk II	Mk III
Curb weight, lb	2150	2480	2520	2530	2650
Wheelbase, in.	90.0	92.0	92.0	92.0	92.0
Track, f/r	48.8/ 50.0	48.8/ 50.0	48.8/ 50.0	48.8/ 50.0	48.8/ 50.0
Length	151.5	158.0	158.0	158.0	158.0
Width	60.0	60.5	60.5	60.5	60.5
Height	49.0	49.0	49.0	50.0	50.0
Fuel capacity, gal.	14.4	14.4	14.4	14.4	14.4
Engine type	ohv 4-cyl	ohv 6-cyl	ohv 6-cyl	ohv 6-cyl	ohv 6-cyl
Bore x stroke, in.	3.44 x 4.38	3.12 x 3.50	3.28 x 3.50	3.28 x 3.50	3.28 x 3.50
Displacement, cc	2660	2639	2912	2912	2912

Instrument panel is little changed in early 3000.

TYPICAL ASKING PRICES

Year & Type	Price Range
1953–1956 100-4	$1800–3000
1957–1959 100-6	$1600–3000
1960–1961 3000	$1600–3000
1962–1963 3000 Mk II	$2000–3000
1964–1967 3000 Mk III	$2800–4000

Prices generally lower in the southwest (southern California) because of greater availability. Cars in the north central and northeast portions of the U.S. and in Canada have often suffered damage from salt used on winter roads.

Wood veneer facia set later 3000s apart.

PERFORMANCE DATA
From Contemporary Tires

	1954 100-4	1957 100-6	1959 3000	1963 Mk II	1965 Mk III
0–60 mph, sec	11.7	12.2	9.8	11.2	9.8
Standing ¼ mi, sec	17.9	18.2	17.1	17.6	17.4
Avg fuel consumption, mpg	23.5	20.0	20.0	18.5	18.5
Road test date	7-54	1-57	1-59	11-62	2-65

regional meets and other happenings increases by the year.

The Austin-Healey is one of the few vintage sports cars still economically available to the average enthusiast. If this big, fast and good-looking machine fits your requirements, you would do well to not hesitate much longer. As the existing supply of cars shrinks, prices of the better examples are definitely on the rise, with no foreseeable end in sight. Good luck and happy Healeying!—*Dave Ramstad*

Driving Impressions: THREE AUSTIN-HEALEYS

ONE OF THE pleasures of authoring a Used Car Classic is the opportunity to drive a good example of the car being covered. In this case, I had the fun of driving three Austin-Healeys at Orange County International Raceway one afternoon, thanks to the efforts of Don Mollett, President of the Austin-Healey Association of Southern California. Don brought his 1960 3000, which he bought in 1971 for a mere $300. It was in sad shape then, but Don has restored the car from the ground up. It is now a very good-looking car and is in good mechanical condition with the exception of the synchromesh.

In addition, two other members of the club brought their cars: a 1956 100-4 owned by Don Rould of Buena Park, California and a 1967 3000 Mark III owned by Jim Whitehead of Fullerton, California. Rould is the second owner of his 100, purchasing it in 1959 and keeping it in excellent shape ever since. Whitehead bought his 3000 in 1972 and repainted and reupholstered the car, making it like new. All three owners were kind (and courageous) enough to allow me to drive their cars and they all managed to hide their fears rather well.

I started out with Rould's black 1956 100 which brought back a flood of memories of my brother's Healey. The clutch is mechanical rather than hydraulic, requiring a deft touch. Once in motion, the gear change to 2nd will also produce a clunk if you're not careful, but this problem soon recedes as you drive along with the top down the wind in your face (literally if you are my size and protrude above the windshield) and that great exhaust note ringing in your ears. The tailpipe exits the left side of the car, just behind the door and ahead of the rear wheel, so you can listen to the exhaust note at all times!

I had remembered my youthful days at the wheel of the Healey as being ones of excitement at the relative power and performance of the 4-cylinder engine. I was happy to note my memory had not glamorized the effect and the power is really there. The brakes are not as efficient as we have come to expect with modern cars, but to hell with the brakes and press on regardless!

I rejoined the group and drove Mollett's 3000 next. The 6-cylinder engine has a more throaty rumble than the four and Mollett's car has that peculiar Healey exhaust whine. Trying to keep in mind the marginal synchros, I sped off down the course, enjoying the power of the six and especially the low-end torque that allows it to pull smoothly from about 2000 rpm up. The brakes felt somewhat better as they should, and the handling was absolutely classic in the sports car tradition, with slight understeer that changes to oversteer with heavy throttle application in a corner. The solid rear axle produces some jumping on uneven surfaces and the ride is firm in the manner of sports cars of that day. None of these comments should detract from the fact that the car handles quite well, bearing in mind the surface of the road, and the ride is certainly pleasant enough.

Whitehead's 3000 Mark III is representative of the last year of the Big Healey. It has the comparatively luxurious features such as the rollup windows and walnut veneer facia of a modern sports or GT car and really does not seem dated much. There is plenty of power, but it seems to be slightly flatter than the earlier 3000. On the other hand, the servo-assisted disc brakes make themselves evident immediately and give the driver the secure feeling that he can use the power and still avoid unpleasantries. Another nice feature of the later Healey is that you don't get the constant flow of warm air on your feet you get with the earlier models. All in all, the Mark III compares quite favorably with most of the present sports and GT cars still on the market in looks, comfort and performance.

Driving those Austin-Healeys brought back a lot of memories and helped to bring into focus what being a car enthusiast means. A small, open car with room for two persons, ample power and a feeling of freedom. These are cars for escaping everyday living and entering the fictional world of wine, women and song and a general state of excitement . . . if only for a few hours on a sunny afternoon.—*Thos L. Bryant*

ROAD & TRACK ON PORSCHE 1972-1975
Road Tests – Owner Survey – New Model Reports – Model Updates – Comparison Tests – Track Tests – European Test – 50,000 M Report – 910 – 911 – 911E – 911S – 914/4 – 914/6 – 912 2.0 – 916 – Carrera RS – 917-10 – Carrera – RSR – Turbo Carrera

ROAD & TRACK ON PORSCHE 1975-1978
Road Tests – New Model Reports – Works Visits – Driving Impressions – Used Car Classic Report – Technical Analysis – Salon – Comparison Tests – 904 GTS Carrera – 911S – 911SC – Turbo Carrera – 914.20 – 917K – 924 – 928 – Speedster Replica

ROAD & TRACK ON PORSCHE 1979-1982
Road Tests – New Model Reports – Owner Surveys – Used Car Classic Report – Model Updates – Comparison Tests – History – Profile – 4WD – 911 – 911SC – 914 – 914/6 – 917 – 924 – 924S – 924S Turbo – 924 Supercharged – 928 – 944

Austin-Healey

THE AUSTIN HEALEY 100 100/6 3000 SPECIALISTS.

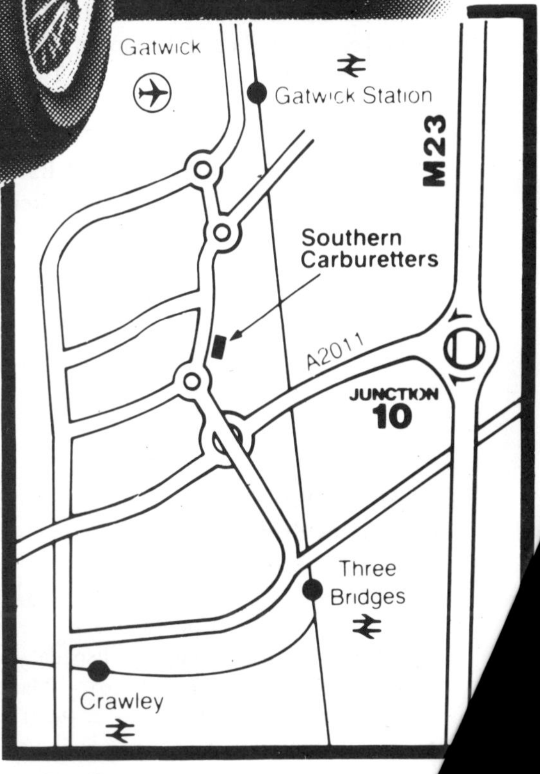

We are proud to be a BL Heritage Approved Manufacturer and Supplier of Austin Healey parts. We have full export/mail order facilities and telephone orders are accepted on credit cards.

Telephone: (0293) 547841/2

SOUTHERN CARBURETTERS

BL Heritage Approved Supplier

Austin Healey Parts & Carburetter Sales

Unit 14, Oakwood Industrial Park, Gatwick Road, Crawley, Sussex RH10 2AZ Telephone: 0293 547841/2

24 hour international order-phone: 0293 546570

R&T

FERRARI & LAMBORGHINI

ROAD & TRACK ON FERRARI 1968-1974
Road Tests – New Model Reports – Racing Classic Road Tests – Owner Surveys – Salon – Comparison Tests – European Tests – Track Tests – 330 GTS – P4 – 365 GTB/4 – 321P – 340 Mexico Berlinetta – 365 GT 2+2 – 250 GT Lusso – 330 GTS Targa – 312B – 246 GT – 265 GTC4 – 166 MM – GTO – Dino 308 GT4

ROAD & TRACK ON FERRARI 1975-1981
Road Tests – Styling – European Tests – New Model Reports – Analysis – Track Tests – Salons – Comparison Tests – 308 GT4 – 308 GTB – 308 GTS – 308 GTi – 166 Inter – 340 Coupé – F1 312 B3 – 512 BB – 365 GT4 – Berlinetta Boxer

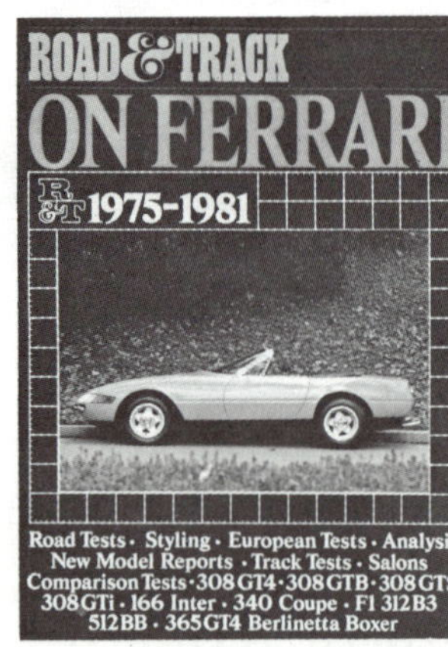

ROAD & TRACK ON LAMBORGHINI 1964-1982
...oad Tests – New Model Reports –
...ifications – European Tests – Styling –
...ison Tests – Countach S –
... – Furia – Jalpa – Espada –
...0 – P111 – Islero – Marzàl –
... S – Roadster – P400 –
...+2

New catalogue available